TAKE A TECHNOWALK

To Learn about Materials & Structures

by Peter Williams & Saryl Jacobson

Edited by Julie E. Czerneda

Trifolium Books Inc.
Toronto

Trifolium Books Inc.
238 Davenport Road, Suite 28
Toronto, Ontario, Canada M5R 1J6

© 1997 Trifolium Books Inc.

Canadian Cataloguing in Publication Data

Williams, Peter, 1942-

    Take a technowalk: To learn about materials & structures

(Teaching innovations)

Includes bibliographical references.

ISBN 1-895579-76-7

1. Materials - Study and teaching (Elementary).
2. Structural engineering - Study and teaching (Elementary). 3. Interdisciplinary approach in education.
I. Jacobson, Saryl   II. Title.   III. Series

TA404.W54 1997   620.1'1'07   C96-931509-0

Project editor: Julie Czerneda
Design, layout, graphics: Roger Czerneda
Project coordinator, proofreader: Diane Klim
Production coordinator: Francine Geraci
Cover Design: Blanche Hamil

Printed and bound in Canada

10 9 8 7 6 5 4 3 2 1

Trifolium's Books may be purchased in bulk for educational, business, or promotional use. For information, please write: Special Sales, Trifolium Books Inc., 238 Davenport Road, Suite 28, Toronto, Ontario, Canada M5R 1J6

This book's text stock contains more than 50% recycled paper.

What's New?

If you would like to know about other Trifolium resources, please visit our Web Site at:

www.pubcouncil.ca/trifolium

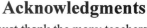

Acknowledgments

We must thank the many teachers who provided us with their thoughtful reviews of this material early in its development: David Gould, Durham Board of Education, Dr. Ann Marie Hill, Queen's University, Marilynn Pascale, City of York Board of Education, Geoff Shilleto, London Board of Education, Derek Tolhurst, Renfrew County Board of Education, Paul Woloszanskyj, City of York Board of Education, Maitland Monro, Scarborough Board of Education, Ted Deuling, St. Laurence High School, Eric Sayle, Simcoe County Board of Education, Nola McIntyre, Sprucecourt School, Nadine Curroro, Hillcrest Education Centre.

    The Publisher also acknowledges with gratitude the generous support of the government of Canada's Book Publishing Industry Development Program (BPIDP) in the production of this book.

    The Authors would like to make special mention of the efforts of Trudy Rising, Mary Kay Winter, Jonathan Bocknek, and Julie Czerneda in developing this project.

                  Thank you all.

Safety: The activities in this book are safe when carried out in an organized, structured setting. Please ensure you provide to your students specific information about the safety routines used in your school. It is, of course, critical to assess your students' level of responsibility in determining which materials and tools to allow them to use.

Note: If you are not completely familiar with the safety requirements for the use of specialized equipment, please consult with the appropriate specialty teacher(s) before allowing use by students. As well, please make sure that your students know where all the safety equipment is, and how to use it. The publisher and authors can accept no responsibility for any damage caused or sustained by use or misuse of ideas or materials mentioned in this book.

Contents

Meet the Authors

Peter Williams

Peter Williams was educated in England where he obtained a B.Sc. in Chemistry. Following a family tradition, he went into teaching, starting his career in a school just outside of London. Canada called, and he spent three years teaching science in Toronto schools. Australia was next on the list and he spent two years there teaching in Melbourne and the small town of Wodonga.

Peter continued his first love of science teaching upon returning to Toronto, moving from high school to elementary school and then into a consultant position. Currently, he is the Coordinator of Science for the Toronto Board of Education. "My responsibilities include science education from kindergarten through high school," he notes.

A published author of a highly successful elementary science textbook series, a children's science activities book, *Light Magic*, and many teacher resource documents at the local and provincial levels, Peter continues to promote his passion of science and technology education for all students. "My latest book, *Take a Technowalk*, reflects my strong belief in a hands-on approach to education."

Saryl Jacobson

Saryl grew up in a house in Montreal that came equipped, she recalls, "with less than a hammer, a five-in-one screwdriver, and a few nails." She was determined to learn how to do things for herself and so entered the Faculty of Education in Industrial Arts and English. That happened to be the last year that someone could enroll in an Industrial Arts program (now called Technology Education program) without a five-year technical background. "At that time, this change effectively ruled out almost every female applying," Saryl remembers. "Since shop was traditionally (and, in most cases, exclusively), a subject only males could take."

Saryl refined her skills and successfully taught shop for 15 years. Her students ranged from adults at night school, to teenagers in secondary school ("including a year teaching machine shop in Inuvik, Northwest Territories"), and to 11 and 12 year olds at the grade 7/8 level.

Continuing in her determination to do things for herself, Saryl also bought into a woodworking co-operative. "I'm a self-employed cabinetmaker," she says with justifiable pride.

Saryl was promoted to the position of Mathematics, Science, and Technology Consultant for the Toronto Board of Education for two years, before becoming a Vice-Principal of a large urban elementary school.

Saryl has written math, science, and technology curriculum for the Toronto and Metropolitan Toronto Boards of Education. Her collaboration with Peter Williams on *Take a Technowalk* has been particularly satisfying. "Peter and I share the pleasure of exploring technology with our students. This book is our way of helping other teachers experience that same enjoyment and success."

INTRODUCTION

What is a TECHNOWALK? Who can go on one? What do you do? It is exactly what it sounds like — a walk with your students through your community to explore the technology that you find there. You have probably taken your students on nature walks around the school, to look at the natural environment. You may also have taken them on longer trips, to a museum, an historical site, a zoo, or industrial plant. Afterwards, students return to the classroom bubbling over with enthusiasm, full of ideas and questions, working together, willing and eager to follow up with further projects. A TECHNOWALK, looking for common materials or particular kinds of structures, can open up a whole new world of discovery and invention.

TECHNOWALKS provide a wonderful excuse to step outside, stretch your legs, and breathe some fresh air. Each time you venture outside, you will focus on something new. Actually, not really new, but new in terms of how you will be looking at it. You will find yourself developing an awareness of technology that you probably did not have before. By looking at only one part of something at a time, separating each part from the whole, you and your students will gain a new ability to analyze our technological world.

So, take a TECHNOWALK with your students. Leave the classroom for an hour, and see what you can find in your local community. There is no need to schedule a field trip months in advance, no need to wait for a bus, no need to rush through a curriculum topic or to try to rekindle interest in something that was completed weeks ago. TECHNOWALKS can be spontaneous events, an extension of a lesson in language or history or mathematics or science, or a way of leading into topics in any of those same or other subjects. Whatever ways you decide to use TECHNOWALKS within your curriculum, we are certain that you and your students will enjoy the investigating, designing, testing, trouble-shooting, and problem-solving that the technological process provides.

What is a TECHNOWALK?

In this book, we have provided ideas for TECHNOWALKS that we have used with our students ranging in age from six to fourteen years old.

What You'll Find in This Book

Find key points here!

Look for helpful tips, suggestions, and directions to reproducible forms here!

Technotalks are ready-to-use forms.

Watch for safety issues, information on advances in technology, and general teaching suggestions here!

Within Each Section

Each section, Introduction, Materials, and Structures, starts with background information to make you aware of the concepts being covered, the recommended approaches to use, and evaluation strategies.

The Materials and Structures sections each contain plans for a sequence of TECHNOWALKS. For each TECHNOWALK, we have made suggestions for Before the Walk, During the Walk and After the Walk, and have included *Technotalk* **Worksheets** that you may want to reproduce for your students. We have also provided extension/home activities. (Get your community involved; they'll love it too!)

The TECHNOWALKS

Each TECHNOWALK is presented with pre- and post-walk activities, each dealing with a particular aspect of technology. They are organized for your convenience as follows:

Before the Walk

These activities serve both as introduction to the technological concept being developed in the TECHNOWALK and to elicit your students' current understanding. You will also find suggestions on how to stimulate interest. Particular safety and planning issues are dealt with as well as the type of record-keeping recommended.

During the Walk

The walk itself -- whether outside, within the school, or within the classroom -- is outlined here. In each case, direction is provided to help students focus on the concept involved. If there are collections or other records to be made, there will be additional instructions here.

After the Walk

It is vital to give students the opportunity to discuss and reflect their findings from excursions such as TECHNOWALKS. You will find suggestions and activities under this heading to help you and your students derive the most benefit from your observations.

Extension Activities

There are additional activities and projects provided with each TECHNOWALK that you can use to extend or enrich the treatment of any particular topic. Some of these are especially suited to older students who may be already familiar with the concepts addressed and need further stimulation.

Kids know a lot about technology. They enjoy taking things apart. They design and build as part of their play long before learning it has a name: technology. They become excited by the adventure of finding out how things work -- and why familiar objects look the way they do.

It is this thrill of discovery that makes the treatment of technology in this book so special. Teachers who have used TECHNOWALKS have found their students carrying that thrill over into the classroom, making it much easier to deliver both concepts and process skills.

Science and technology

It is worth taking a moment to think about the distinction between these two human endeavors. Many people think of technology as applied science, and it certainly can be. However, people used technology long before they understood the science behind the processes they were using. Egyptian wall paintings, for example, show metal workers at a copper smelter, in about 1450 B.C, about 2000 years before the discovery of the chemical process involved.

Basically, science aims to increase our understanding and ability to explain nature, while technology aims to design and develop devices and processes for practical purposes. For example, science explores how gravity and friction act, but technology has given us the wheel.

Traditionally, the teaching of technology has been associated with machines and workshops. Today, it includes computers and information technology. Another trend is the movement of technology education from the workshop into other subject classrooms.

This has been a movement that has enhanced the teaching of other subjects. The problem-solving and discovery inherent in design and technology are appropriate across the curriculum. And, since wherever there are human beings, there is technology, learning more about the world means learning more about technology. That it's fun too is a bonus!

What is Technology?

You will find a form to help you elicit your students' ideas about technology on page 77 of the Appendix. A sample reaction:

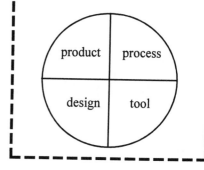

Technology can be as simple as a pair of scissors, as innovative as skates for pavement, or as sophisticated as a shuttle capable of operating in space and gliding through the air.

Making Sense of Technology

The parts that make up a technological device — materials, structures, machines, power and energy, and systems/control — are quite often right at hand and easily taken for granted. Yet most of us have asked, at some time or another these questions.

The first two questions are the focus of this book. However, all of these questions are part of the consideration of technology which you and your students will undergo during your TECHNOWALKS.

What is it made of?	material
How is it built?	structure/fabrication
What is it used for?	function
How does it work?	mechanism
What makes it go?	energy and power
What regulates it?	systems/control
How does it look?	aesthetics
How efficient is it?	ergonomics

Materials

The exploration of materials -- what things are made of -- is used to introduce technology in this book. By focussing on this aspect first, students are able to recognize these important relationships between function and composition. Elicit ideas from students about the following:

Natural or Synthetic?
Before the 1930s, clothing was made from plant and/or animal fibres — cotton, linen, or silk. Then chemists discovered how to make synthetic fibres from chemicals in coal, oil and natural gas. The new material — nylon — was stronger and lasted longer than the natural fibres.

Different materials may be combined.

Objects serving the same function may be made from different materials.

Each material has characteristics that produce a certain effect in a structure.

Downhill Slide!
For many objects, wood has been supplanted by new materials. You rarely see wooden sleds or toboggans in stores or catalogues today, but they were the only kind available a few decades ago. The next material commonly used was metal. Today you see children sliding down hills using sheets of plastic or even reused cardboard packaging!

Structures

Students will take what they have learned about materials and technology from the first TECHNOWALKS and apply this learning to structures -- built objects -- in the second half. Students will learn, through observation and their own construction, more about the relationship between function and composition. They will explore the essentials of design as they test and explore their own assumptions.

A structure may support a load. For example, a stool is designed to have someone sitting on it.

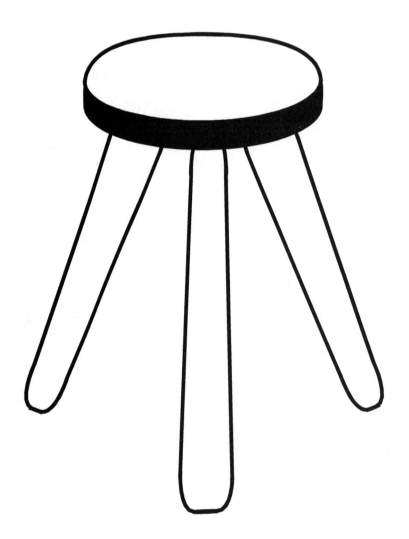

A structure may span a gap. For example, a bridge may be used to span the distance between two banks of a river.

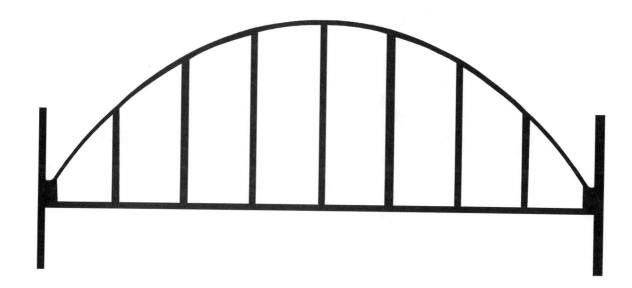

A structure may enclose objects or people. For example, a school encloses students, teachers, and the materials they use every day.

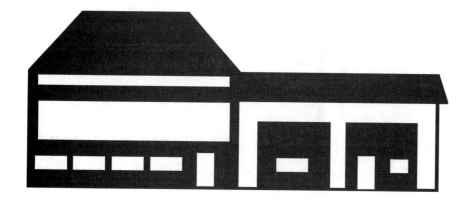

Within the classroom, you can illustrate the relationship between material and function using students' backpacks or lunch bags. Discuss with students if they would make a backpack out of paper. How about a shopping bag out of metal? Look at a backpack and record the many different materials you will find contained in that one object.

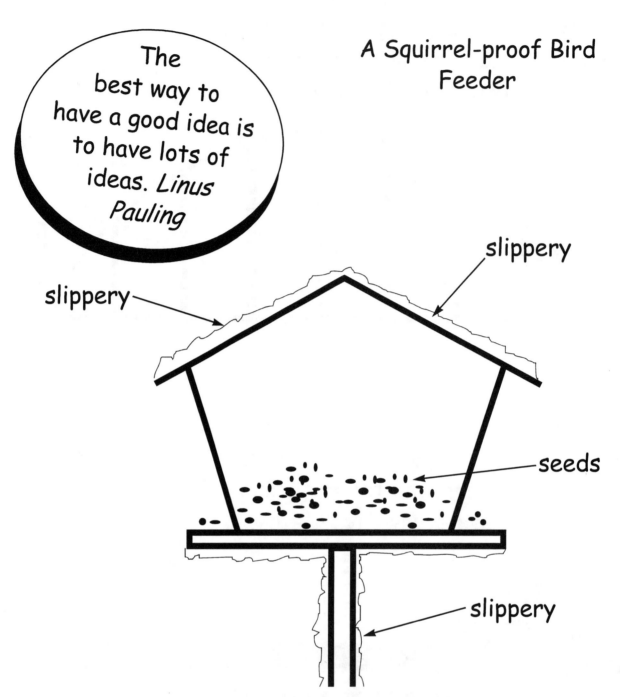

The best way to have a good idea is to have lots of ideas. *Linus Pauling*

A Squirrel-proof Bird Feeder

slippery

slippery

seeds

slippery

A squirrel would not be able to jump on my bird feeder and eat the seeds because I made the roof and sides very slippery. To do this I put lots of shortening on top of the wood. *Gale*

A Squirrel-proof Bird Feeder

Students will need time to get used to the idea that there is no one "right" answer in technology. Every approach has value and the best approach can only be determined after testing. Once they are comfortable with this idea, you will find creative solutions such as these beginning to fly around the room!

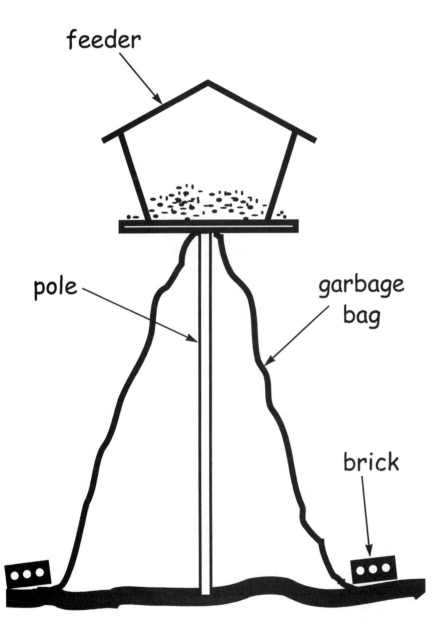

feeder

pole

garbage bag

brick

I saw that squirrels were climbing up on our bird feeder from the ground. To stop them from doing this, I put big garbage bags around the pole. I used bricks to hold down the edges of the bags. *Gino*

You will find detailed outcomes for each TECHNOWALK in the Appendix, pages 72-3.

Encourage students to persevere. It is always easier to discard something that is not going well, but to stick with it and revise it until it works most often results in something far superior to what one first considered.
Help your students recognize that making changes does not necessarily constitute failure or poor designs; change will likely result in improvements in the overall design.

Outcomes: What do We Want Students to Gain?

Technology encourages students to learn through doing and, in the process, to become self-directed learners. This results from both independent and interdependent learning. Technology encourages collaborative work, engaging the students in discussion, sharing and consideration of others. So much of what emerges is the result of people working together to plan, design, and create a product. It is often not the actual product that is where real learning occurs, but rather in the process of working together to come to a common end result.

The following lists the general outcomes to be expected from introducing technology to your students with TECHNOWALKS.

Students will

- understand how the form, shape, color, texture, strength, and structure of a thing relate to its function and purpose.
- create and use models and pictures
- understand the need for orderly procedures in group work and be able to develop such procedures in co-operation with others.
- be able to find more than one solution to a problem, and respect other people's solutions.
- be able to safely use simple tools and materials and to build simple objects and models.
- be able to make things using a variety of materials, tools, and methods
- be able to design and make useful objects out of waste materials and evaluate and improve them.
- be able to identify and use the most appropriate and attractive materials for specific purposes.
- be able to explain the connections between the way people live, technology, and the environment.
- be able to describe the features of their neighborhood and the issues that affect it.

Problem-solving

An education in technology will provide students with opportunities to explore open-ended problem solving. They will learn to recognize the design procedure as a sequence for developing and completing projects in a self-directed manner. They will learn to recognize the design procedure as a problem-solving method.

Curriculum Delivery Using
TECHNOWALKS

Ages 6 to 8

Where to start? The primary focus from ages 6 to 8 should be to help expand children's knowledge of technology. You'll find this most easily done by providing technological experiences both inside and outside the school.

Promote the growth and development of practical communication, reasoning, and interpersonal skills through group explorations. For example, students might work together to plan and build a birdhouse or feeder for the school yard. At the next Parent-Teacher Association meeting, student representatives from the class might attend the meeting to show off the results. Students might even go so far as to record the types of birds that the birdhouse attracts.

> Expand their knowledge of technology.

Ages 9 to 11

From ages 9 to 11, you can expand students' explorations into other subject areas and contexts: school, family, community, and the environment. Encourage the use of correct technological terminology. Your students should begin to recognize these concepts across the curriculum and beyond. This is the appropriate age to start using a design process to make plans and to construct products.

To help your students with the technological vocabulary, supply opportunities for these terms to be used throughout the classroom. For instance, as they build, ask questions and help them express what they are doing using wider vocabulary. Post the terms you want used in the classroom, and illustrate these through drawings or photographs cut out of magazines or newspapers. Some can be written up on cards with definitions to explain their meaning.

> Link technology to other subjects and contexts.

Some of the vocabulary you can help your students develop through their explorations are listed in the *Glossary*, pages 74-6 of the Appendix.

If your students are participating in technology workshops at the local high school, incorporate this experience into your TECHNOWALKS. For example, student observations on structures can be woven into a workshop project.

Ages 12 to 14

When working with these students, expect them to continue to use the design process for a variety of purposes. If you are unfamiliar with the design process as used in technology, it is not mysterious. It is simply a sequence of steps to arrive at a successful conclusion.

The **first** step in a design process is to identify and define the problem. The **second** step involves a whirl of ideas as all of the possibilities are considered. The **third** step is to select the best possibility or solution from all of the choices. Once a selection has been made, the next steps are followed in sequence, from making a plan or design, carrying out the plan or making the design, testing the plan or design and, finally, making any necessary modifications to ensure success.

Your students will begin to acquire specific skills related to design, such as using the proper tools and machines, drawing systems (including computer-assisted drawing programs), and creating production plans. Community TECHNOWALKS will be particularly useful to help you show students the connections between the design and technological concepts learned at school and how they apply outside the school environment.

Students use the design process to arrive at successful solutions.

Using this Book

TECHNOWALKS are short, highly focussed outings, which shouldn't require much advance notice. Standard field-trip forms, signed by parents/guardians at the start of each school year, are generally all that you'll need in terms of permission to take your students outside on short walks. You may need to line up, in advance, parents who could be available on short notice. Safety is also a consideration and, in particular, there should always be the proper ratio of adults to students.

A letter to parents, outlining the nature and purpose of the TECHNOWALKS, is an important preliminary action. It should be sent home at the start of the year so that parents are aware of the focus of your program.

Choosing the Right TECHNOWALK

We have provided ideas for 10 TECHNOWALKS in this book. You will notice that they begin with materials (what something is made of), and move to structures (how something is formed). Each new TECHNOWALK builds naturally on the others, so we recommend that you proceed in order.

Can some TECHNOWALKS be left out? Concepts explored in one are necessarily touched again in subsequent activities. Depending on your students' experience with technology, you could pass over one or more that might already be familiar to them from other programs.

When to Use TECHNOWALKS

On page 20, you will find a summary of the TECHNOWALKS in this book with suggestions on when and how to use each one. While you may find in your school that some TECHNOWALKS are better suited for K-3 or 4-6 students, many teachers have told us that each of the TECHNOWALKS is very suitable for 7-8 students, especially those who have not yet been exposed to either these activities or to technology education in general.

The TECHNOWALKS in this book have been deliberately designed so minimal advance preparation is needed. You do not, for example, need to arrange a bus or, in most cases, to schedule your visits to specific locations. So sharpen your powers of observation, decide which adventure you and your students shall undertake, and get TECHNOWALKING!

Preparing for TECHNOWALKS

Technotalk memo for friends and parents on page 22. Information and permissions letter on page 87.

TECHNOWALK Components

Alert students and their parents that the walk can take place in a variety of weather conditions and to dress appropriately.
Make sure that all students have provided a signed permission form. You will find a form for your use on page 87 of the Appendix.

Before the Walk

This is the time to brainstorm with your students about what they know of the technology concept to be explored and what they expect to observe regarding this on the walk. This information can be recorded on a flip chart or chalkboard so that the students can refer to it later to check on their learning.

This is also a time for teaching about the concept as necessary to clear up any misunderstandings or major gaps in the students' understanding.

Your students can use this time to develop their own recording sheet to take with them on the walk. The students might consider other innovative but appropriate ways of recording the walk, such as a tape recorder or camera.

Find out what your students already know about the technological concept to be covered during the walk.

Find out what your students expect to find during the walk.

During the Walk

Each TECHNOWALK involves moving and observing, however not all are excursions outside. Some can take place inside the school or even within a classroom.

Here the students will be actively involved in observing the technology concept "in action." Encourage your students to look for all aspects of the concept and to record their observations. Help them through your questions and directions. Be prepared to hone your own powers of observation!

Question and direct their observations.

Plan ahead. Will you need volunteers?

Consider all safety issues and be prepared.

After the Walk

This is the time to consolidate the walk. The students can check their recordings against those they predicted in Before the Walk. Discussion can take place to come to a common understanding about the concept observed. Clarification can be given where there is some disagreement about what was seen and how it fits into the technology concept.

Extension activities are included to further consolidate the students' understanding of the concept. These can be in the form of research assignments, hands-on activities to model the concept studied, or open-ended problem-solving activities to challenge the students to incorporate the technology concept into their solution of the problem.

We recommend that your students work in groups during all aspects of the TECHNOWALK. This allows them to support each other through discussion. The group will also allow students to remind each other of the safety aspects on the walk.

Evaluation and Assessment

How do you know if your students have gained knowledge and skills from the incorporation of TECHNOWALKS in your program? We have provided several ideas and tools to help you find out, including:

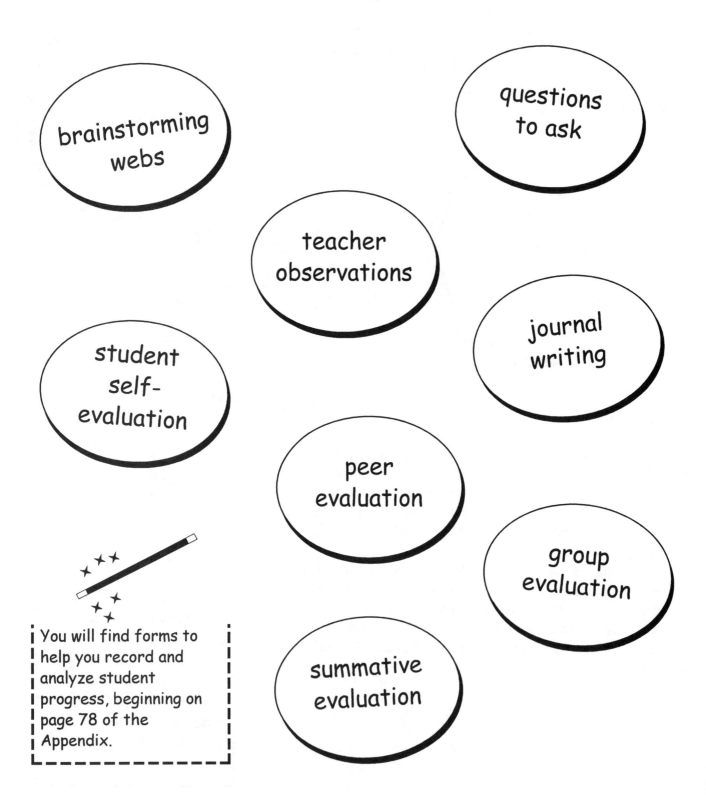

brainstorming webs

questions to ask

teacher observations

student self-evaluation

journal writing

peer evaluation

group evaluation

summative evaluation

You will find forms to help you record and analyze student progress, beginning on page 78 of the Appendix.

Student Brainstorming Web

Every TECHNOWALK begins with an elicitation activity. If you wish, have students put down their ideas in a web. Asking for a second web later in the technology portion of the class for comparison will help you identify the learning that has taken place. This is now a summative form of evaluation. For example, these webs show one student's growth in understanding of the concept of Materials.

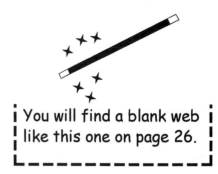

You will find a blank web like this one on page 26.

Before the Walk

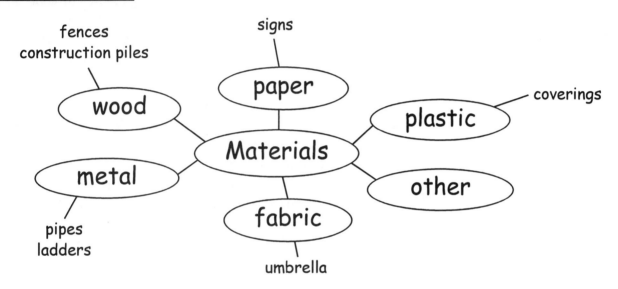

After the Walk

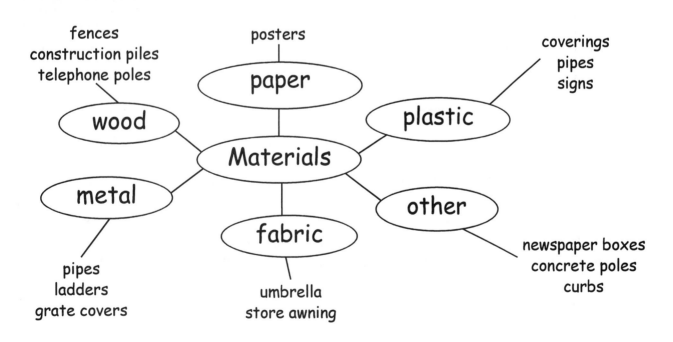

You will find forms to help you determine the success of group work and to help students assess their own performance, starting on page 80 of the Appendix.

Questions

After each TECHNOWALK, a series of questions is included to determine the level of thinking and understanding of the concepts that are developed during the TECHNOWALK.

Observations

By listening to your students as they talk to each other and to you, by looking at their recordings, and by observing them during the activities, you can assess their growth and work in technological concepts.

We recommend that you keep ongoing records of your observations together with samples of each student's work. In this way, you can identify a student's ability to work with others, interests, and areas of weakness, progress, and success.

Journals and Portfolios

Have students write in a **journal or log** about their accomplishments through the different TECHNOWALKS. They can use this for planning, recording, and reflecting on what they have done. Teachers can also record comments here.

Student self-evaluation is an important part of the evaluation process. Encourage the students to develop a **portfolio/folder** in which they can record the activities they have done, discuss why they liked them, and save samples of their work. This will encourage the students' articulation of their feelings, attitudes, and values with respect to technology learning.

Teamwork

Because the TECHNOWALKS involve students in groups or paired, you may wish to use the group self-assessment form supplied in this book to help you determine how effectively the students felt they worked together.

Summative Evaluation

Summative evaluation, based on Bloom's Taxonomy, can be used at the end of each section to evaluate the students' knowledge of technology, their understanding of the technological concepts, and their skills in problem-solving. An overview that you can use to help devise evaluation questions is provided on the next page.

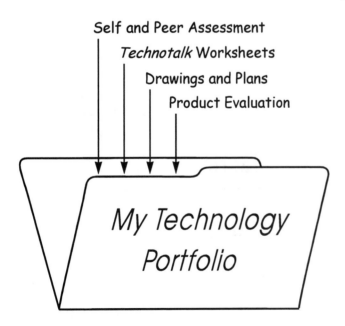

Self and Peer Assessment
Technotalk Worksheets
Drawings and Plans
Product Evaluation

My Technology Portfolio

Overview of Summative Evaluation

knowledge

describe, label, identify, name, state, locate, list, define, outline.

e.g., Describe five different types of building materials.

About Technology

comprehension

explain, give examples, summarize, rewrite, paraphrase, convert, distinguish, predict.

e.g., Explain, with examples, why some structures are stronger than others.

analysis

analyze, break down, differentiate, discriminate, illustrate, identify, outline, point out, select, separate, sub-divide, categorize, classify, distinguish.

e.g., List at least three types of balls used for sports and classify them by their materials and function.

Guided Technology

application

infer, change, discover, operate, predict, relate, show, solve, use, design, build, manipulate, modify, demonstrate, compute.

e.g., Design and build a structure to shelter a small animal.

evaluation

judge, compare, contrast, evaluate, criticize, justify, draw conclusions.

e.g., You are going to build a bridge from found materials to support a specific number of toy cars. Justify your choice of materials and design.

Problem-solving & Inquiry in Technology

synthesis

devise, compile, design, compose, explain, organize, rearrange, plan, combine, categorize, show relationship, synthesize.

e.g., Compose and write a short story that imagines the lives of the first family to live in the oldest home in your community.

Overview of the TECHNOWALKS

TECHNOWALK	Summary
LEARNING ABOUT MATERIALS	
1: Classifying Materials	Students examine properties of various materials.
2: Checking the Bounce	Students explore how different materials are used.
3: Staying Together	Students examine how materials are fastened together.
4: Playground Adventure	Students observe and examine the materials and fasteners used in playground equipment.
5: Perspectives	Students learn the effect of perspective on how they observe.
LEARNING ABOUT STRUCTURES	
6: Classifying Structures	Students use their observations to classify structures by function. They test the forms that supply strength to structures.
7: Towers	Students observe and test how materials are assembled in order to produce towers.
8: Going Over	Students examine bridges in the context of design, materials, and function.
9: In or Out	Students observe fences and other types of enclosures, then design and make their own.
10: The Building Business	Students observe work at a construction site, looking at technology and related careers.

Integration Opportunities
Use *TECHNOWALK 1* as an introduction to a science unit on properties of matter.
You could also use this TECHNOWALK as part of a history unit on early settlers by repeating the activity in a museum or pioneer village.

You will find Detailed Outcomes for each TECHNOWALK on pages 72-3 of the Appendix.

TECHNOWALK 8: If you do not have bridges of any type in your area, you can substitute a "virtual" TECHNOWALK for your students using film, video, and other resources.

Preparation | Where to Walk

Preparation	Where to Walk
sample materials; recording sheets	Community or classroom
assortment of sports balls; samples of rubber, elastic, etc.	School yard or gymnasium
sample fasteners; glue-making supplies	Community and school (*optional* at-home component)
model-making materials	Local playground
drawing materials (*optional* cameras, etc)	Community
found materials; popsicle sticks; glue & other fasteners	Community and School
found materials; decks of playing cards	Community
construction materials	Community
found materials	Community
permissions and safety plans	Community

Integration Opportunities

Use *TECHNOWALK 2* as part of a unit in mathematics or science on presenting information in graph form.

Use *TECHNOWALK 8* as part of a unit on medieval times. Modify by having students build a model castle with a bridge and moat.

TECHNOWALK 10: If you are unable to arrange a visit for your class to a construction or renovation site, consider combining the use of video resources with an in-class visit by a general contractor who can answer questions about technology and careers.

Technotalk

For our Technology Cupboard, we are collecting:

found materials such as:
   small unused scraps of wood
   popsicle sticks
   plastic bottles
   fabric
   buttons
   laces and strings
   crayons
   film canisters
simple tools such as:
   screwdrivers
   hammers
   hand saws
   pliers
   glueguns
other supplies such as:
   duct tape
   masking tape
   white glue
   carpenter's glue
   paper clips
   decks of playing cards
   elastic bands
   velcro strips
   plasticine
   sandpaper
   nails, screws, bolts
   toothpicks

Dear Parents and Friends

Our class is getting ready to start an exploration of the world around us. We will be taking TECHNOWALKS to learn more about the materials and structures that are part of our lives. After each walk, we will return to the classroom for hands-on activities in technology.

You can help!

Join us on our walks!

Share your knowledge & experience!

Help fill our Technology Cupboard!

Please check:

○ Yes, please contact me about joining the class on a walk.

○ Yes, please contact me about coming to a class to share my knowledge and experience.

○ Yes, please contact me about donating materials for your technology cupboard.

Name: _____

Phone Number: _____

Preferred time to be contacted:

_____

Thank you!

Learning About Materials

When we look around us, a moving panorama expands before our eyes. Within this wider scope are countless images, each different and each contributing to the overall picture. If we choose to, we can focus our eyes on particular objects - a house, a car, a lamp post, a tree. If we start filtering this perspective into an even narrower, more specific image by restricting our field of vision to only one aspect of that object, we can start isolating the separate concepts of technology.

The Concept

In this section, we are going to focus on one concept of technology: materials. When looking at human-made objects; i.e. any technology, consider some of these questions:

- What material is used?
- Is there more than one material used?
- Why is a particular material(s) used?
- What would happen if another material were used instead?
- How does the choice of materials affect the use?
- How does it affect the durability? Appearance?

Students should be encouraged to play and work with a wide range of different materials. They can build with wooden blocks and commercial plastic joining materials; create arts and crafts using paper products, plasticine and wire; and model using toothpicks, popsicle sticks, and glue.

The Technology Cupboard

Even if your school has a Technology Facility or other area set aside for the tools and materials your students will need, we recommend you maintain a Technology Cupboard in your classroom. This will allow you and your students to easily and quickly do hands-on activities both as part of the TECHNOWALKS and as part of your other programs.

Some ideas of what to include are listed on the letter to parents on the opposite page. Use clear or labeled containers to help keep your supplies in order. You should also contact parents before they send in bags of useful items. This type of donation can accumulate more quickly than you may be able to use it.

For Safety:
- Make sure cords are removed from old appliances that students may be taking apart.
- Avoid toy tools. Sharp, well-made tools are safest.
- Discard pieces of scrap wood that contain nails, glue, or finishes.

You will find a list of books which explore the idea of materials in particular, on page 84 of the Appendix. You might wish to work together with your school librarian, to discuss what the class is working on, to share your book list, and to ask for further suggestions.

Literature and Technology

An important role of the study of technology is to make students aware of technology as the basis of many everyday things they encounter. Literature can help them make this connection in a familiar and comfortable manner.

Through books, your students can begin to explore the many links between literature and technology. What immediately comes to mind is the story of *The Three Little Pigs* who build their houses with straw, sticks, and bricks. The image of the wolf huffing and puffing and blowing the house down provides a clear, visual impression that children can readily understand.

Literature is an integral part of the TECHNOWALKS in this book, especially when used as before or after activities. Use literature as a basis for research as well as for an enjoyable "hook."

Consider having your students write their own stories that incorporate elements from a TECHNOWALK. For example, a story could be developed around a material or structure that students enjoyed discovering. Encourage them to imbed what they have learned into their creative train of thought.

Before the Walk

Brainstorm Activity

Generate a list of different materials with your students. Include only what something is made of, not considerations such as shape, design, color, purpose, and size. Ask these questions.

> What is meant by the term material?

> What materials are used to make the objects in our classroom, outside in our community, and in our homes?

Technotalk Worksheet

- Have students record the materials they know in the form of a web. A *Technotalk* form is provided on the next page. Repeating this web after the TECHNOWALK will provide an indication of what the student has gained.

Optional: Collecting and Classifying Different Materials in the Classroom

If your students are unfamiliar with identifying materials and their properties, have a "dry run" by having them make a collection from objects in the classroom (or at home). The time taken here will make their work during the TECHNOWALK much easier and more rewarding.

1. Ask students to collect at least ten different objects. They may, if the object is too large, choose to draw or sketch it instead.
2. Once they have a sizeable collection of objects and drawings, ask them to classify each object according to type of material (wood, paper, plastic, metal, fabric, or other).
3. Next, have students perform a further classification of the materials, this time according to the properties of the material. You may wish to provide samples of materials that illustrate the major properties. There is a *Technotalk* form that you and students can use for a properties checklist on page 88 of the Appendix.

TECHNOWALK

1
CLASSIFYING MATERIALS

hard/soft

shiny/dull

stiff/bends

rough/smooth

Where to Walk
Outside: schoolyard, one or two quiet residential streets, area of small stores.
Inside: lumber or hardware store, art gallery, museum

How Long to Walk
Outside: 1/2 hour
Inside: 1/2 to 1 hour

Technotalk

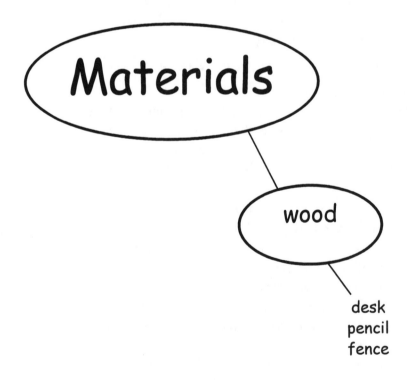

CLASSIFYING
MATERIALS

Make a web to show all of the different types of materials you know. Add examples of each.

Materials

wood

desk
pencil
fence

Technotalk

Name: _____

CLASSIFYING MATERIALS

Use checkmarks to record which type(s) of material are in each object. Use a question mark if you are not certain.

Object	Type(s) of Material in the Object					
	wood	*paper*	*plastic*	*metal*	*fabric*	*other*

You will find two sample student webs on page 17.

To help students organize their information, there is a properties of material checklist on page 88 of the Appendix.

During the Walk

If your students did the optional classroom activity, challenge them by asking them not to duplicate any objects already collected when they go outside.

1. Have students work outside in pairs. Each pair will need an empty bag and paper/pencil to sketch objects that cannot be contained in the bag or removed from their location.

2. Student pairs will collect as many different objects as possible, either as samples or drawings.

After the Walk

Technotalk Worksheet

Students can use the worksheet provided on page 27 to help them record the materials that made up each object collected.

1. Ask students to select one single property that best describes the material of a particular object. They should be prepared to justify their selection.

For example, a rock would be classified under Type as *Other*. Its properties are: *hard, rigid, heavy, inelastic, smooth/rough (depending on the rock), and probably dull*. The one single property that probably best describes it is *hard*. However, other properties could be justified.

2. Discuss with students their observations of the different materials they noticed on the TECHNOWALK.

Why are certain materials used in some structures and not in others?

How might an object be affected if one of its materials were exchanged for another? (Consider a brick car. A plastic house.)

3. Once the students have identified one property for each material, as a class complete a chart such as the one following outlining the material, property, and function. (At first, limit students to one function, then go back over each property and add additional functions.)

Sample Chart

Material	Property	Function
wood	strong	building
glass	transparent	windows
fabric	soft	clothes

Research Suggestions (Individual or group)

- What is a synthetic? What was the first synthetic material produced? Who made it and what was it first used to make?
- Find out about a modern material, such as nylon, mylar, teflon, kevlar, or velcro. How was the material developed? Who invented it? What are its uses?
- What are different types of glass? Why does some glass break and some shatter? What makes glass shatterproof? What makes glass non-glare? tinted?
- Look at bicycles or barbeques or cars left outdoors over a period of time. Discuss the effects of rust on metal. What can you do to inhibit rust?
- What are plastics? How are they formed? What are their uses?

Extension Activities

More about Types of Materials

1. Divide the students into groups to investigate one of the following materials: wood, paper, plastic, metal, or fabric. Students from each group record everything they can find in their own homes that is made up of that particular material. For example, the list from a "wood" student might include floors, tables, chairs, doors, mixing spoons, cutting boards.
2. Give students time to combine their findings with other members of their group.
3. As a class, prepare a graph to answer this question: *What is the most common material in most homes?*

More about Properties of Materials: Strength & Transparency

1. There are many ways students can test the strength of materials.
- For solid materials, place a piece of the material across a gap between two tables and determine the mass that can be supported by the material before it bends or breaks.
- Test threads for tensile strength as follows: use 30 cm lengths of different threads; e.g., string, wool, yarn, word, dental floss, hair. Span a stick across two surfaces, tie one end of a thread to the stick and the other to the handle of a bucket or plastic

Want to stimulate some discussion? Bring in the following pairs of objects and ask students how they are related (the second is made from material similar to the first!)

- machine oil, pantyhose (both from petrochemicals)
- machine oil, plastic toy (both from petrochemicals)
- clear plastic pop bottle, polar fleece garment (garment made from recycled plastic of bottle)
- newspaper, facial tissue (tissue made from recycled paper)

Materials must be strong enough to withstand the stresses that are placed in a structure during normal use. The strength of a material must be suited to the structure in which it is used.

Brainstorm materials to wrap (insulate) a jar of warm water to help keep it warm longer -- or to keep an ice cube cold!

Interview an older person. Share the results.

container. Add standard masses until the thread breaks. Order the threads according to their tensile strengths.

2. Students can investigate transparency by collecting, sorting, and making a display of objects that are transparent, semi-transparent, or opaque.
- What makes some objects more transparent than others?
- What is the difference between glass and plastic?
- What happens to glass or plastic when the surface gets scratched?

Create a Collage

Using drawings and sketches or photographs from newspapers and magazines, make collages to show the different types of materials found on a TECHNOWALK, specifically one for each of the following: wood, paper, plastic, metal, fabric.

What Was? What's Here? What's To Come?

If you wish, discuss as a class what questions should be asked in an interview. You may also wish to focus on a specific period of time.
- Interview an older person to find out about household objects which are now made from different materials.
- Set up a table listing objects whose materials have changed over time. Discuss the advent of electricity as it affects these objects.
- Make predictions about how you foresee the material of these objects changing again.

Sample Chart

Object	Made from Then	Made from Now	Prediction
meat grinder	wood	metal	?
iron	cast iron	plastic/teflon	?
washer	wood board	metal machine	?
cupboards	wood	plastic laminate	?

Before the Walk

1. Discuss with students what makes a ball fun to play with or good for a sport.

Are there any balls that are not used in a sport or game? What is their function?

What type of balls are not round?

2. Elicit from the class what material is used to make most balls (elastic, rubber). Discuss what it is about these materials that makes them a good choice for this purpose.

During the Walk

Take a TECHNOWALK to the gymnasium and see how many different kinds of balls students can collect. (For example, tennis ball, football, baseball, soccerball, volleyball, beach ball, pingpong ball, etc.) Bring as many as possible back to the classroom (or do the *After the Walk* activity in the gymnasium).

After the Walk

1. Students will investigate the bounciness of the balls they collected.

Technotalk Worksheets

Go over the recording sheet and graph paper on the next two pages with your students to ensure they understand what they need to do.

2. Have students work in groups of three. One student will drop the ball from a height of 1 m, another will count the bounces until the ball stops, and the third will record the results. Students can record their findings on the *Technotalk* worksheet provided on page 32.

3. Repeat the activity, this time measuring the height of the first bounce.

2

TECHNOWALK

CHECKING THE BOUNCE

baseball

soccer

golf

football

tennis

pingpong

Where to Walk
Outside: community center with sports facility.
Inside: gymnasium
  How Long to Walk
Collecting the balls should take about 15 min. Testing the balls for their bounce, etc., should take about 1 hour.

Technotalk

Checking the Bounce

Record your results on this chart.

Type of Ball	Number of Bounces	Height of First Bounce

Technotalk

Checking the Bounce

Graph your results of the number of bounces on this sheet. Make a similar graph to show your results for the height of the first bounce.

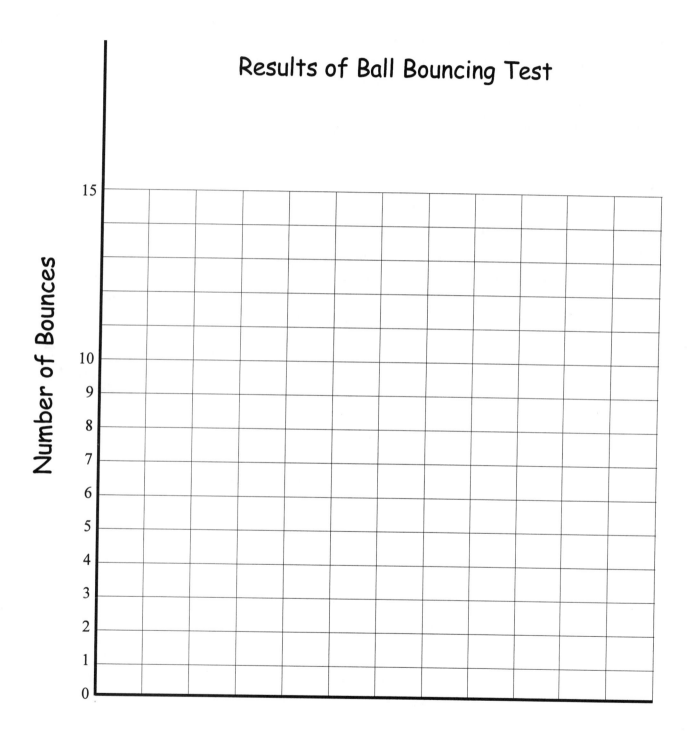

Results of Ball Bouncing Test

Number of Bounces

Type of Ball

No Mistake!
Researchers at General Electric thought they had failed in their attempt to produce a useful synthetic to replace rubber. What they had made seemed useless. But in 1947, Peter Hodgson bought the material, hired a student to make small balls from it, and the result, "Silly Putty," was an instant commercial success.

4. Allow each group time to graph their results for the ***number of bounces*** test on the *Technotalk* Worksheet, page 33. Then, ask each group to make a list of the balls tested in order of ones that bounce the least (or not at all) to the one that bounces the most number of times.

5. Have groups then graph their results for the ***height of the first bounce*** test. Again, ask them to summarize their results in a list that orders the balls tested from least to most.

6. Go over the results as a class. Ask these questions.

Is there a common characteristic or feature among those balls which bounce most and which bounce highest?

What properties make the material of different balls more suitable to one sport or game than another?

Extension Activities

Rolling Distance
Another way to test the balls would be to determine which ball travels farthest along the ground when rolled. Someone can use chalk to mark where the ball comes to a stop.

Classifying by Properties
Compare the properties of different manufactured balls. Order the collection of balls according to different properties, such as soft/hard, flexible/rigid, light/heavy, smooth/rough.

Going Internal
Segment the balls to see which ones are hollowed out, which ones are filled. (We recommend that you prepare one set of sample balls to use.) What are the materials inside? Are they different from the ones outside? How does the material affect the function?

Flotation
Test the balls in water to determine which float or sink. What is the correlation between buoyancy and the material used?

Before the Walk

1. Elicit from your students what kinds of things need to be held together. Decide on a working definition of the term *fastener*.

2. Send students looking around the classroom to collect as many examples of fasteners as they can in 2 min. When they are back in their seats, generate a class list. (A typical list might start with items such as velcro, buttons, laces, nails, screws, glue, tape, rubber bands, string, thread, thumb tacks, paper clips, staples, etc. Point out fasteners students might miss such as hinges, different kinds of tape, clips, snaps, and so on.)

3. Ask these questions.

> What does each fastener hold together?

> What material(s) are used to make the fastener?

4. Ask students how many ways they could organize this list. For example, would they classify fasteners by the materials they hold together?

Technotalk Worksheet

Your students can use the *Technotalk* form on page 36 to help them organize their information about fasteners.

During the Walk

1. Students will make a list of the fasteners found outside on a TECHNOWALK. (The list might include hinges, nails, bolts, screws, rope.) If you wish, supply a second copy of the *Technotalk* form on page 36 or have students design their own form to use.

2. Students should note what each fastener holds together. Do any fasteners work on more than one type of material?

3

STAYING TOGETHER

T E C H N O W A L K

buttons & bows

nuts & bolts

hinges & hooks

Hooked!
Velcro, named after the French *velours* for velvet and *crochet* for hook, was invented in 1948 by a Swiss engineer who noticed that burdock seeds stuck to his socks by tiny hooks.

Where to Walk
Outside: any quiet area with homes, shops, or small business.
Inside: in the school, visit a hardware or craft & hobby store, a mall.

How Long to Walk
Outside: 1/2 hour
Inside: 1/2 to 1 hour

Technotalk

Staying Together

Record what you find on this chart.

Fastener	Where I Found It	What It Holds Together

Technotalk

Staying Together

Here's a recipe for your own super glue!

Ingredients

- ○ skim milk
- ○ vinegar
- ○ water
- ○ baking soda
- ○ strainer
- ○ pot
- ○ jar
- ○ stove
- ○ popsicle sticks

Directions

1. Mix 250 mL (1 cup) of the skim milk with 45 mL (1/4 cup) of the vinegar in a pot.
2. Slowly heat the mixture on the stove.
3. Stir until the mixture thickens and then remove the pot from the stove. Continue to stir as long as lumps are forming.
4. Separate the lumps from the liquid using the strainer. Throw away the liquid and put the lumps into a jar.
5. Add 30 mL (2 tablespoons) of water and 2.5 mL (1/2 teaspoon) of baking soda to the lumps.
6. When all the bubbling has stopped, you will end up with a jarful of very sticky white glue.

What Next?

- ○ Test your glue by using it to fasten two popsicle sticks together. Once the glue has dried, try to pull the two sticks apart. How strong is your glue?
- ○ Use your glue to make as tall a structure as you can out of popsicle sticks. Test the strength of your structure by placing weights on top of it.
- ○ Change the recipe! Experiment! For example, add food coloring, sparkles, or pieces of crayon to your glue.

Post-it-Notes

The story of Post-it-Notes began in the 3M Corporation during the 1960's. The corporation was trying to find a superpowerful new glue, and Spence Silver was working on the project. Silver was a chemist with a playful streak. He sometimes changed the chemical "recipes" he was working on, adding a bit more of this and a bit less of that, just to see what would happen. One day he ended up with a glue that would stick two things together, but then you could pull them apart again. A perfect glue to make a piece of removable notepaper!

How
do you know
which fastener to
use?

Doing a unit on
magnetism? Develop
thinking skills about
materials using these
activities:

- conduct experiments
  to find out which
  materials allow
  magnetism to pass,
  such as wood, paper,
  cloth, metal, water,
  or plastic.
- create a game or toy
  that uses a magnet.

After the Walk

1. Discuss the list of fasteners found outside. How do these compare with fasteners used in the classroom?
2. Make a list of fasteners found in and around the house. (The list might include zippers, shoelaces, buttons, snaps.) Do any fasteners work on more than one type of material?

Technotalk Worksheet

The *Technotalk* worksheet on page 37 contains a recipe and directions for making a homemade glue. The glue is actually casein glue. To turn milk into glue, first the milk must spoil. This happens when bacteria in the milk produce acid. To speed up this effect, students add vinegar (an acid) and heat. The spoiled milk has two main ingredients: curds (lumps) and whey (liquid). Both of these substances are in regular milk, but the acid makes them separate. The curds are made up of the protein casein. When baking soda is added to the casein, a chemical reaction occurs to produce the glue.

Extension Activities

More about Fasteners

1. Have students use as many different household fasteners as they like to create a cloth puzzle.
2. Make a class wall hanging that shows how fasteners work.

Shoe Patrol

Students can investigate the types of fastenings used in shoes. To integrate with mathematics, have them remove and measure laces, then graph the length of lace against the shoe wearer's height.

Holding Power

Students can test the ability of different fasteners to do their job. An excellent enrichment activity is to have students compare brand name glues and how well they live up to their advertised ability.

Puzzle Power

Students could design a simple puzzle without any connecting parts and challenge a friend to complete it. Start with a square, and make one straight cut in any direction, then a second, etc. How easy is such a puzzle to solve compared with a puzzle that stays "fastened" together?

Before the Walk

Brainstorm with students the different structures/ apparatus found in a playground, including swings, see-saws, slides, climbers, sandboxes. Ask these questions.

> What is the most popular part of the playground? Why?

> What materials are used to make each structure?

> How are the different parts fastened together?

During the Walk

1. On a TECHNOWALK in your community, look at the different structures/apparatus in a playground.
2. Have students record the materials and fasteners used. Encourage them to sketch and label each structure.

Technotalk Worksheet

Use either or both of the *Technotalk* forms on the following pages to help record observations.

4

PLAYGROUND ADVENTURE

TECHNOWALK

swings

ramps

ladders

slides

Where to Walk
Outside: any nearby playground facility, either at the school or a local park.
Inside: some shopping malls, restaurants, or hotels have indoor playground equipment. Check first if it will be worthwhile visiting.
  How Long to Walk
Outside: 1 hour
Inside: 1 hour
(Allow time to "test" the equipment.)

Technotalk

Name: _____

Playground Adventure

Record your observations on this chart.

Equipment	Material	Fastened with

Technotalk

Name: _____

Playground Adventure

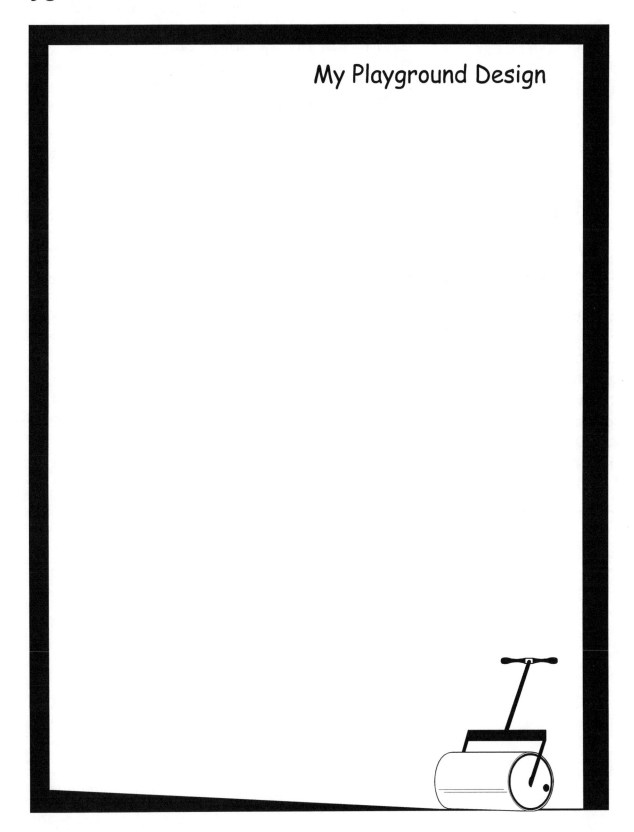

My Playground Design

After the Walk

1. What were the predominant materials found in the playground?
2. What materials were used the least?
3. Have students make a model of one piece of equipment in the playground, using only one type of material. Consider that fastening devices will also have to be made of the same material. Ask students to justify their choice of material.

Extension Activities

Suited to the Job

Combine class data to make a class graph of the materials used in playground construction. Which materials are most often used? Find out more about why these materials are selected.

Personal Playgrounds

Return to this activity after students are familiar with the basic forms used in structures (TECHNOWALK #6 plus any of 7-9). Have students design their own personal playground. The project could be done simply as a drawing, or taken to a model stage using found materials or construction kits.

Food Frenzy

Challenge students to make a moving piece of playground equipment using only one kind of food (material) and one kind of fastener. They could try marshmallows, jelly or bread cubes, fruit slices, etc. If you ensure a clean working area, this activity could be done as part of lunch.

Who can play? Encourage your students to think about the accessibility of playground equipment. Could a child who cannot climb stairs or ladders still have fun at your local playground? What changes might be made to help everyone enjoy it?

Before the Walk

This walk is a fun way to help students develop their observing and recording skills. It is also a good introduction to drawing techniques that they can use now and later. At the same time, students gain an insight into how the *way* they see the world can affect how they interpret what they see. Start discussion of these ideas using the *Technotalk* form on the following page.

Technotalk Worksheet

- Show students the illustrations on the worksheet and ask what they see first. Promote discussion about why some people pick out different images.

- Discuss what perspective means. Have students look at objects from different perspectives (above, below, across). For example, have students sit under their desks and think about how objects in the room look. What do they notice about a person walking past?

> How can you tell if something is far away from you?

> When you look at something across the room, how can you tell how big it is?

During the Walk

1. Explore different ground surfaces - stone, rock, gravel, tar, concrete, brick, grass, dirt, etc. Encourage students to make rubbings of different textures. Sketch patterns in patios and fences. Look for lines of symmetry.
2. Standing on the ground, look up and record a list of the different structures that are visible overhead (including wires, poles, towers, planes, kites, roofs).
3. From the same location, look straight out and record the different structures visible.
4. Repeat, this time looking down.

Optional

If you wish, combine this activity with an art project in perspective. Have students portray some technology they observed during their walk as part of their work.

5

PERSPECTIVES

TECHNOWALK

up/down

in front/behind

under/over

smaller/larger

far/near

Where to Walk
Outside: any location that gives students a safe place to stand and draw, as well as do rubbings, such as a park, the schoolyard, or other location.
Inside: it is preferable to do this outside, but any large open area, such as the gymnasium can be a good start.
How Long to Walk
Outside/Inside: allow at least an hour for sketches.

Name: _____

What do you see?

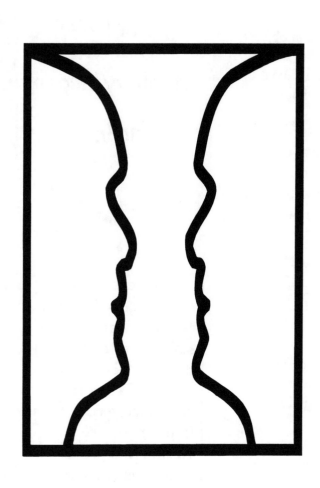

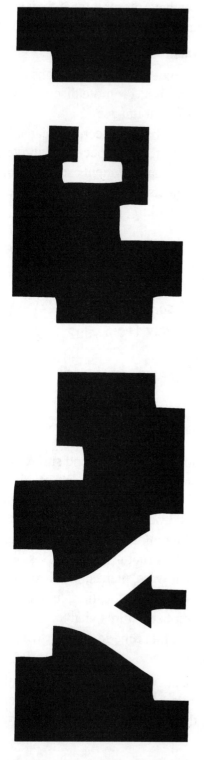

Technotalk

Perspectives

Draw another car in this box that looks closer than the car in the first box.

Draw another car in this box that looks farther away than the car in the first box.

Draw another car in this box that looks the same distance away as the car in the first box.

After the Walk

1. Sketch a familiar landscape from memory, trying to get the shape and proportions right.
2. Invent a geometric landscape constructed from lines, planes, shapes and solids.
3. Choose either sketch to do one finished perspective drawing (a technique for representing three-dimensional objects on a two-dimensional surface).

Extension Activities

Looking Down

If possible, take a trip up an elevator in a tall building to explore the perspective of ground coverings when looking down from a height. (As an alternative, show a video, slides, or drawings of such a view.) What are the materials seen on rooftops? What do the actual ground coverings look like from a distance? Look down on a rainy day and sketch what you see. (One student's pictures were filled with circles representing umbrellas!)

Magazine Magic

Have students start a class collection of magazine advertisements that use perspective to give the illusion of size or unusual location to certain objects.

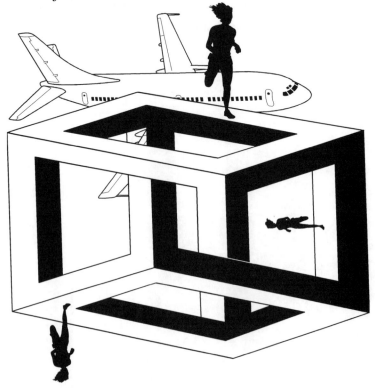

Drawing perspective allows the artist to readily convey the illusion of height and distance. However, when younger students look at images like this, they may be fooled at first into thinking that the shapes are rendered accurately. Simple horizon exercises, in which lines converge at a point, can be used to help them understand the difference.

Learning About Structures

As in the previous section on materials, it is important to narrow the focus of students to the specific concepts being considered. For this reason, the term **structure**, as used here and in much of the literature of technology, refers specifically to a built, supporting framework.

A structure is a built, supporting framework.

This framework has three main functions.

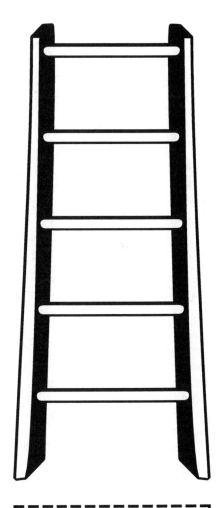

The Concept

In this section, we expand the investigation of materials into the consideration of another concept in technology: structures. It helps students to classify structures by their function because this leads to a consideration of how the materials, fasteners, and shapes used help each structure perform its function.

Structures have different functions, including:

- to support a load (e.g. a ladder)
- to span a gap (e.g. a bridge — which could be something as simple as a fallen tree or as complex as a large, commercially constructed bridge)
- to enclose objects or people (e.g. a car or building)

Structures can serve more than one of these functions at a time.

- a tower could support a tank (a load) that encloses something (water).
- a building could have one or more floors that extend over a gap.

The Technology Cupboard

Check and see if you need more supplies before starting these TECHNOWALKS. You may want to have construction kits available as well.

> There can be other functions beyond those listed. For example, an important function of many structures is to be beautiful or to communicate an idea.

Structural Concepts

In order for a structure to function properly, it must resist forces acting on it and not collapse. There are two types of forces that act on a structure: one is static (not moving); the other is dynamic (moving). Static forces are less destructive than dynamic forces. A structure must be strong enough to resist both the internal and external forces that act on it and one way that it can gain or lose strength or flexibility is by changing the shape of materials.

As students begin to explore structures, they will develop an understanding of structural concepts, including:

bending/buckling	too much force on a structure (i.e. too great a mass) will cause it to bend or buckle in the middle.
folding	folding material in half lengthwise will double its thickness but not necessarily make it more rigid. Opening up the folded material to create an L or right angle will result in greater rigidity.
corrugation	involves repeated folding of materials. The rigidity is increased only in the direction at right angles to the fold.
tension/compression	tension is a stress that acts to pull a material apart; compression is a stress that acts to squash a material.
shearing/torsion	shearing refers to cutting; torsion refers to twisting both ends of an object equally in opposite directions.
cellular forms	cells and tubes resist pressure applied to the ends, particularly under compression (e.g. a coiled paper roll is stronger from end to end than in the middle)

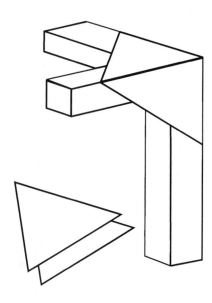

A simple method of joining wood pieces that yields excellent results in building model structures is to use glue and cardboard corners. Use this method after students have experimented with the strength of different geometric shapes.

A lighthouse has strong foundations to withstand strong winds and huge waves. It is rounded in shape so that the wind and waves will flow around it more easily.

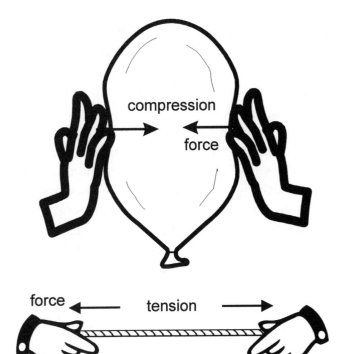

compression force

force ← tension →

Students can experiment with the effect of compression or tension on their structures. While they should test one at a time, you may want to mention that both of these forces typically affect a structure. For example, the seat and legs of a chair are being compressed when someone sits in it, and any cross-pieces connecting the chair legs will be under tension.

force

force

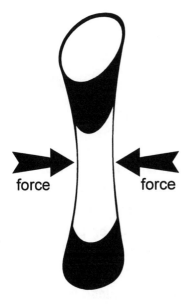

force   force

Using boxboard cylinders such as toilet paper rolls, you can readily demonstrate how a shape can have more strength in one direction that another. Groups of cylinders can be used to provide strength to resist force applied to the ends.

Lego® Houses
An innovative contractor in Cobourg, Ontario, was recently asked to help build 200 000 new homes in Japan. This contractor no longer uses traditional wood-framing techniques. Instead, he uses Styrofoam "lego-blocks" that are stacked in the desired shape, then filled with concrete. The Styrofoam stays in place as insulation. Because of the strength of the concrete, and the way the Styrofoam can be stacked, walls can easily curved or made at unusual angles.

You will find a list of books which explore the idea of structures in particular, on page 85 of the Appendix. With your school librarian, discuss what the class is working on, share your book list, and ask for further suggestions.

Structures and Materials

Materials vary in their ability to resist stresses. It is therefore important to know the strengths of different materials when building structures for different purposes. Each of these structural concepts can be demonstrated by students to other students.

Structures and Design

Through an exploration of structures, students will develop an awareness that simple shapes or elements — squares, circles, triangles — may be the basis for many structures. Students can research unusual structures, including: towers, domes, castles, silos, barns, pyramids, lighthouses, igloos, tepees.

Some parts of structures need to be a certain shape to work properly. Others need not necessarily be the shape they are. Students can work in pairs to analyze the shape of different structures.

> What are the most common shapes for support?

> What is the strongest shape?

Design and Function

1. Discuss the function of different common-place objects.
2. Pass an object around and see how many different functions one can imagine for that particular object (e.g., a pen is used for writing but might also be used as a straight-edge, a drumstick, a baton). Students will have a more difficult time suggesting alternative functions for structures that are fixed, such as doors and windows.
3. Some objects are quite different yet both have one function. Make a list of different objects that share a common function. List the similarities and differences of both. For example: a letter and a telephone are both used to communicate. They are similar in that both involve using words to transmit a message. Also, both can involve great distances. They are different in that a letter is written and sent by mail, while a telephone involves a spoken message that is sent through wires etc.

Before the Walk

With your students, make a list of terms that they can use to describe a structure. Here are some starter questions to ask students to get them headed in the right direction:

What is a structure?

How do people (or other living things) use structures?

What different kinds of structures are there?

6
CLASSIFYING STRUCTURES

TECHNOWALK

scaffold

building

fence

ramp

bridge

During the Walk

1. Have students look for as many different structures as they can, recording their findings as notes or sketches. (You may wish to take along a videocamera to make "visual notes.")

Technotalk Worksheet

Use the *Technotalk* form on the next page to help students record their observations. Encourage the use of sketches as well.

2. For younger students, divide them into groups each assigned to find structures that either:
    *   support a load,
    *   span a gap,
    *   or enclose objects or people.

    Older students can record their observations of function for each structure.

3. Students should observe the ways in which the structures are built.

Where to Walk
Outside: around your school & neighborhood; any quiet residential street.
Inside: in a mall that has displays and structures.
How Long to Walk
Outside: 1/2 to 1 hour
Inside: 1/2 to 1 hour

Technotalk

Name: _____

Classifying Structures

Record your observations on this chart.

Structure	Materials Used	Function(s)

Technotalk

Classifying Structures

Write the name of each structure you observed in the box that best describes its function.

supports a load	spans a gap
encloses objects or people	**other**

53

After the Walk

1. Discuss the students' findings from the TECHNOWALK.

Technotalk Worksheet

Use the *Technotalk* classification form to help students group their structures by function. Use a similar chart to collect class results.

2. Compare the structures that support a load, span a gap or enclose objects or people. What are the similarities?
3. Do any structures serve more than one purpose?

Extension Activities

The Strength of Shape

Use straws or pipecleaners to make a strong frame. It must not change its shape when a force is added.

1. Make different two-dimensional shapes (triangles, squares, pentagons, etc.) from pipecleaners.
2. Test the shape to find out how rigid it is by laying it flat on a desk and gently pulling on opposite end joints. Repeat the test by pressing gently on the opposite end joints.
3. If necessary, make the shape more rigid by adding or removing pipecleaners.
4. Based on their observations, identify the shape that could be used to make the strongest structure.

Stability

- Create a structural design that supports a mass.
- Determine what structure uses the least amount of material and still maintains its strength.
- Design a structure that will reach out as far as possible from a fixed point and remain stable.

Card Challenge I

Challenge students to build the tallest structure possible using playing cards. How many cards were used?

- Try to build the tower again, to the same height, using fewer cards.

Card Challenge II

Challenge the students to build a structure that:

- supports a load (measure its strength);
- spans a gap (measure the length it spans);
- encloses objects or people (determine the capacity/volume).

> A triangle is the only shape that cannot be altered by pressing or pushing on its joints.

Before the Walk

Towers are structures that share at least one function in common -- to be tall. They capture students' imagination and are a great topic for construction challenges. Engage students in a discussion about towers. Some questions to ask:

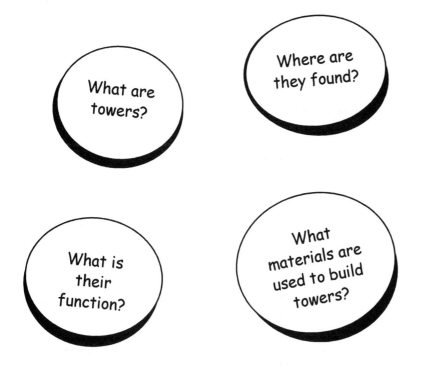

7

TOWERS

TECHNOWALK

skyscrapers

tv towers

telephone poles

signposts

stoplights

During the Walk

1. Take a TECHNOWALK in your community to explore examples of tall structures. Make sketches of each tower and list all the materials that are found in each. (A drive through an industrial area (or farming area) can lead to very interesting towers. If you do this, videotape what you see for students to review.)
2. What are the uses of tall structures?
3. What is the tallest structure in your community?
4. Are there special ways the materials are arranged? Encourage students to make sketches.

Technotalk Worksheet

There is a *Technotalk* form on the next page that students can use to record their observations.

5. How are the building materials fastened together?

Where to Walk
Outside: any streets with sidewalks that pass by towers. (Keep in mind that many things are "towers" to small students.)
Inside: it is preferable to do this walk outside. If you cannot, then consider a virtual walk using video and other resources.
How Long to Walk
Outside: 1/2 hour

Technotalk

Name:_____

Towers

Record your observations in this chart. Make sketches on the back of the page to help you remember the shape of each tower.

Tower	Purpose of Tower	Materials in Tower

Technotalk

Towers

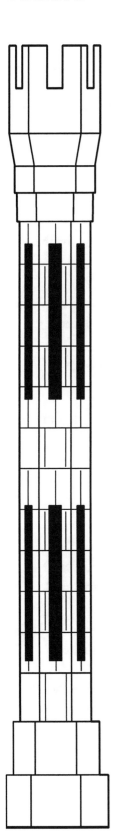

Tower Challenge!

Sketch your design here.

Over the Top!
The 553 m CN Tower in Toronto, Canada is the world's tallest free-standing structure. This is a vital part of its function. By being taller than the surrounding buildings, the CN Tower can be used to broadcast signals for television and radio without interference.

It is recommended that you restrict the amount of found and joining materials to be used. Here are some suggestions:

• drinking straws and paper clips;

• a single sheet of letter-size paper and a 20 cm length of masking tape;

• old newspapers and tape (If newspapers are spread out, they are not very strong. If they are rolled into tubes and taped together, they will be strong enough to support a brick.)

After the Walk

1. Discuss with the students the towers they observed on their TECHNOWALK. What are the common materials used? Ask them to think about similarities and differences between the materials and methods used to make towers and those used to make lower structures.

2. Discuss the function of each kind of tower. You may wish to talk about why towers have warning lights (for aircraft) and why these lights usually blink (as a better warning and to prevent birds from being attracted to the towers).

Extension Activities

Technowalk Worksheet

There is a form students can use to record their designs for these or other tower challenges on page 57.

Tower Challenge I

1. Ask students to build the tallest tower they can using only one type of found material and one type of joining material. Ask students to:

• Explain why they chose a particular material and how successful they were at building the tower.

• Describe the limitations of the building materials.

• In light of their experience, choose another material and build another tall tower.

• Determine whether using a mix of materials might help in building the tallest tower.

• Modify their designs and try new suggestions.

Tower Challenge II

• Use different materials to make the tallest tower.

• Use a different base to increase the height of the tower.

• Build the tallest leaning tower using one type of material.

Tower Challenge III

• Support the mass of a marble placed on top of the tower. If the tower falls over, what changes could the student make to ensure that the tower can support the marble?

• Test the ability of the tower to stand up to the wind generated by a small fan. If it blows over, what changes does the student need to make to ensure that it does not? Try this again with a mass, such as a brick, placed on top of the tower.

8

GOING OVER

TECHNOWALK

overpass

walkway

railbridge

culvert

Before the Walk

1. Pose the following problem to your students (rephrasing as necessary).

A farmer leads her cattle to the pasture every morning and home again every evening. There is a small ditch on the way, but the farmer and cattle can easily jump over it. After a week of rainy weather, water flowing through the ditch washes away the sides. It soon becomes too wide to jump. What can the farmer do?

> What do people use bridges for?

> What materials are used to make bridges?

2. Encourage the students to call out their ideas. Record every one on the board or chart paper. Some will suggest bridges, but others might suggest filling in the ditch or going around it.

3. Elicit ideas of what materials might be available on a farm to make a simple bridge. Ask students how the bridge could be tested without risking either the farmer or cattle falling into the ditch.

Technotalk Worksheet

There is a *Technotalk* form on the next page that you can use to present the scenario.

During the Walk

1. Take a TECHNOWALK to look at examples of bridges in your area. If there are no actual bridges, look at any type of crossing, including ramps, and overhead bridges that connect two buildings. You could also conduct a "virtual" TECHNOWALK using videotapes, pictures, and other resources.

2. Have students make a list and draw sketches of the various types of bridges and crossings that they find.

Where to Walk

Outside: anywhere with bridges, ramps, etc. Be aware of hazards such as traffic, height, or water. Alternative: visit a mini-golf or other area with bridges.

Inside: some buildings contain bridges, but these are not usually easy to observe. Consider a virtual walk using video and other resources. (A local railroad association may have model bridges.)

How Long to Walk

Will vary with what you choose to do.

Technotalk

Going Over

A Farm Situation

These cows and the farmer who cares for them walk each day from the barn to the pasture. One day, their way is blocked by a ditch that has been filled with water by a rainstorm.
The ditch is 2 m wide. Cows like to have a solid surface to walk on that is at least 1 m wide.
Help the cows and farmer reach the pasture without getting wet.

What shapes are strongest?

What materials could you find on a farm?

What type of fastener will work best?

Technotalk

Name:_____

Going Over

Research Record

Type of Bridge:_____

Materials Used:

_____

_____

First Built (when, where, by whom, etc.):

_____

_____

Advantages	Disadvantages

Modern Example:_____

I found this information in:

_____

_____

Provide information to students on the history of bridges. A bulletin board display of pictures can be developed, illustrating the various types.

After the Walk

Discuss the different types of bridges or crossings that were found on the TECHNOWALK.

Technotalk Worksheet

If students are researching the different types of bridges, use the *Technotalk* research form on page 61 to help them organize and compare their information.

Extension Activities

Bridge Challenge I

Challenge students to build a bridge from popsicle sticks and masking tape that will span a predetermined space between two desks and support a toy car. Place the desks so that they are one and a half popsicle lengths apart. (An additional challenge: use 100 toothpicks and a small amount of white glue to build a bridge that would span 30 cm and support an object that represents the mass of an average adult.

Bridge Challenge II

Introduce different styles of bridges and ask the students to construct their own models.

Bridge Challenge III

Students will design and build one style of bridge using only found and joining material available. The bridge must span a gap of 2 m. The structure must support as great a mass as possible.

- How can you make the bridge as strong as possible?
- What happens if you increase the span of the bridge?
- What happens if you increase the mass?
- How does the height affect the design?
- What combinations of materials can be used to increase the strength?
- How can you test the strength of the bridge?

Arch — a bridge supported by one or several arches.
Cantilever — built on beams which rest on cantilevers, or brackets sticking out from the banks. The cantilevers are supported by huge towers.
Suspension — suspended from cables anchored to each bank. A roadway is hung from the cables by means of its two supporting towers.
Beam — supported on land at either end by tall columns or piers.

The Fixed Link
The bridge linking Prince Edward Island and New Brunswick across the Northumberland Strait spans 13 km of ocean in a series of arches. The bridge was built in 175 pieces on land, then each piece was placed in the water within 150 mm of its planned location. The placement was made this accurate by using information from the Global Positioning Satellites.

Before the Walk

One of the functions of structures that students may not have thought about is to keep things in -- or out! Ask students what kinds of things are being kept out of their classroom (weather, noise, dust, other people etc.). Ask them what kinds of things are being kept in (warmth, paper, themselves, etc.).

During the Walk ❝ ⸜ ❝ ⸜ ❝

Technotalk Worksheet

There is a *Technotalk* form on the next page for students to use to record their observations.

1. Have students record the function of each fence they observe. Encourage them to give descriptive names to each that will help them remember.
2. If you wish, have students take along tape measures and record the height and gap of the fences. Before you do this, be sure that there will be no problems with students walking up to fences that may be on private property.
3. Spend some time investigating gates and how they work.
4. Have students rank the fences in terms of aesthetics as well as function.

9

IN OR OUT

privacy

wind

pets/small children

security

T E C H N O W A L K

Where to Walk
Outside: any quiet residential street. An interesting choice can be a local landscaper that may have displays of fencing for students to observe.
Inside: any larger hardware or landscape store may have displays.
How Long to Walk
Outside: 1 hour
Inside: 1/2 to 1 hour

Technotalk

Name: _____

In or Out

Record your observations on this chart.

Fence	Keeps In	Keeps Out	Material(s) Used

Technotalk

In or Out

What needs to be kept inside? Write or draw your ideas on this page.

Longest Fence
Australia's dingo-proof fence is still the world's longest fencing project, consisting of a 2 m high wire fence 5530 km long. It separates wilderness -- and the dingos, wild hunting dogs -- from Australia's main sheep farming areas. The fence is no longer maintained by the government.

Encourage students who are interested in endangered species to find out how fences are being used on islands such as Hawaii to protect rare animals and plants from imported predators such as pigs, goats, and rats.

After the Walk

1. Create a class list of the different types of fences observed and their function.
2. Challenge the class to make a model fence that will be able to contain a small motor-driven or windup toy, such as a car, on a desktop. The fence must also satisfy these parameters (or some other set of parameters of your choice):
- withstand the "wind" from a hand held hair dryer for 1 minute
- look pleasing to the eye
- be made from found materials

Extension Activities

Packaging

1. Packages are technology that also keeps things "in or out." Have students collect different samples of packaging. They can examine the materials used in terms of strengths, stresses that the materials must withstand, unusual shapes/designs, etc.
2. Students can be challenged to find an item that is packaged in different materials in different shapes/designs (e.g., milk is packaged in cardboard boxes, in glass bottles, in metal cans, in plastic bags, in plastic bottles. It also comes in many different sizes, from the small restaurant style containers to large 2 L containers.) What are the advantages/disadvantages of each?

Horse Feathers

In horse jumping, portions of fences are used as obstacles in the course. These fences must stand up but also must fall apart easily if the horse or rider touches them. Have students design their own fence. They could decorate these fences with the school colors or logo, or with an advertisement for a local company (sponsor).

Before the Walk

The building industry is composed of a great number of occupations dealing with different materials and structures. Find out what your students know about careers in this area of technology. Here are some starting points.

Who works on a new building? On other structures?

How do people learn to do these jobs?

10

THE BUILDING BUSINESS

T E C H N O W A L K

apprenticeship

trade

college/university

During the Walk

If there is a building under construction (or demolition) in the area, have the students sketch or make a photographic record of the changes in the structure. Students can learn to differentiate between restoration and renovation. They can explore buildings that are deemed historical and compare styles (materials and structures) to today.

Ask students:

- What materials can you see?
- How are the materials moved?
- What is the foundation made of?
- What safety devices are in place (clothing, transport)?
- What machines are used? What are they used for?
- Who do you see working? What are their jobs called?

Technotalk Worksheet

There is an activity on page 68 that you can use to help students with the vocabulary used in building careers.

Plan well ahead if you will be visiting active construction sites. Much of this work may be seasonal in your location. As well, there are several safety issues to be addressed. You may wish to prepare a videotape of a building under construction for your class instead. Consider inviting a general contractor or building inspector to work with your students.

Technotalk

The Building Business

Any project requires experts in several different areas. If you are interested in the Building Business, you might want to check out all of the possibilities! Here's how to start. Beside each of the following construction stages for a building, write the name(s) of the job(s) involved.

Then find out more about the ones that are new to you.

Here are a list of jobs to get you started: bricklayers, carpenters, delivery drivers, drywall installers and finishers, gas fitters, sheet metal workers, glaziers.

Job to be done	Who does it?
designing the building	_____
surveying the property	_____
excavating the foundation	_____
putting up basement walls	_____
pouring the concrete floor	_____
supplying power	_____
supplying gas	_____
supplying water and sewer lines	_____
supplying telephone and other services	_____
putting up the wooden framework	_____
shingling	_____
insulating	_____
adding brick	_____
installing electrical fixtures	_____
putting in ductwork for heating/cooling system	_____
putting in pipes for water/waste	_____
putting in wiring for phones and cable	_____
installing windows and doors	_____
putting up drywall or panels	_____
plastering	_____
painting	_____
installing wood trim and cabinets	_____
installing flooring and carpets	_____
landscaping	_____
checking the quality of construction	_____

Adapted with permission from the *Career Connections* series
©Trifolium Books Inc. and Weigl Educational Publishers

Technotalk

Name: _____

The Building Business

Here are some questions you might like to ask someone about their career.

1. What do you like about your work?

2. What do you not like about your work?

3. How did you find out about your career?

4. What training or education would I need to do this work?

5. Do you work with other people or by yourself most of the time?

6. What tools and machines do you use?

7. Do you need a computer?

8. What do you think might be the same about your job in 10 years? What might be different?

9. What did you want to be when you were in school?

10. What do you like to do when you are not at work?

Take advantage of the real estate listings shown on the Internet or television as a source of home designs for your students.

After the Walk

- Invite a guest speaker to talk to your class after this TECHNOWALK. An ideal choice would be either a general contractor, or a career specialist from the local high school or college. Be sure this person understands the kind of question your students will be asking.
- It is also very helpful, regardless of the age of the student, to have a guest show the tools and clothing worn on a construction site. This is an ideal introduction to the wide variety of careers that are involved in building.

Technotalk Worksheet

There is a form on page 69 that your students can use to help them interview someone in the building business about their career.

Extension Activities

Houses

Make a list and draw sketches of different types of housing in your community. Included in this list might be bungalows, town houses, apartment buildings, two-story semi-detached houses, ranch-style houses, Victorian houses, split-level houses, etc..

Construction Site

You are responsible for managing a new construction site and require machines for use on that site.Design and construct one or more pieces of construction equipment, such as a crane, ladder, hoist, cement mixer, or truck.

House Building Challenge

Design and build a house that can be put together and taken apart easily, using as many found and joining materials as needed.

Dream House

Design and build your dream house.
- What materials would you use?
- How many rooms would it have?
- How would it be furnished?
- How would it be heated?
- Where would it be built?
- Draw plans and build a model (diorama) of your dream house

Career Exploration

Encourage students to keep an ongoing record in their portfolio or journal of occupations they find interesting. Younger students could begin a scrap or picture book to help them remember what they learn. Older students may want to gather more detailed information and to explore certain careers further in the library, etc.

If any of the students or parent volunteers has a doll house, model railroad, or other scale models of buildings, see if these can be brought into class for the class to examine. A followup activity could be to design a room or addition to the model building.

Appendix

Contents

Detailed Outcomes

Inquiry, problem-solving, decision-making, recording results, data management including collecting, organizing, displaying (graphing), and interpreting data from a variety of sources, design and building using common materials, as well as community exploration are common to all of the TECHNOWALKS.

Technowalk	Mathematics	Science	Technology
1: Classifying Materials	sorting and classifying ordering	classifying by criteria properties of materials	properties of various materials uses/functions of different materials
2: Checking the Bounce	numeration making accurate measurements ordering speed and distance	buoyancy making inferences and predictions decision-making	segmenting safe use of tools evaluation
3: Staying Together	measurement sketching	acids properties of materials strength and force	testing strength of fasteners exploring safe building connectors assembly of materials concrete materials and adhesives
4: Playground Adventure	surface area measurement angles sketching	environmental impact safety properties of materials strength and force	models safety in design simple and complex technologies economic factors
5: Perspectives	symmetry shape and proportion geometry sketching	use of senses perspective and observation communicate solutions, ideas, and procedures in a variety of ways interpretation/prediction	texture perspective drawings recognize and describe patterns in natural and human-made environments models

Technowalk	Mathematics	Science	Technology
6: Classifying Structures	geometry sketching classification	testing strength classification testing stability	building structures explore shapes and materials for constructing prototypes relationships among technology, people, and the environment
7: Towers	linear measurement geometry estimation/comparison symmetry	mass and weight balance and stability gravity testing criteria	design construction building techiques evaluation
8: Going Over	measurement scale drawing	communication properties of materials testing criteria	bridge structures selecting appropriate materials material assembly
9: In or Out	measurement perimeter and area number patterns sketching	wind, velocity strength, stresses properties of materials	models patterns in natural and human-made environments making prototypes
10: The Building Business	estimation organizing surveys	safety interpretation interview skills	careers safe building techiques factors affect building design and construction

Glossary

Aesthetics - the characteristics of an object or system that appeals to the sense of beauty.

Bending - If a force is applied to the centre of a structure such as a plank, it will bend in the middle as the force is increased. The material has to be rigid enough to resist the force and therefore not bend. If a force is applied at the end of a structure, without support, bending will occur at that end.

Cellular Forms - the combining of several tubes to produce a structure that is effective at resisting pressure applied to the ends of the tube, i.e. compression.

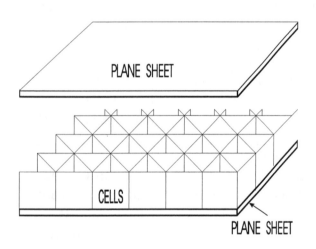

Compression - where a load tends to squash an object, such as when a person leans on the back of a chair.

Control - the means by which mechanisms are regulated.

Corrugation - the repeated folding of material to provide rigidity at right angles to the folds and flexibility in the direction of the folds.

Demand - characteristics or features that must be addressed in the design solution. For example, safety is a demand. (Engineering terminology refers to demands as specifications.)

Design - a proposed solution to a problem.

Design brief - a statement of the problem, giving demands and preferences.

Device - something designed to serve a particular function.

Efficient - able to do something with little waste of time or energy.

Effort - a force applied to a machine to produce an action.

Ergonomics - the efficiency of an object or system in relation to the work performed by the human body and mind in using it.

Evaluation - how well the design meets the essential demands and preferences.

Exploded view - a drawing that shows an object as if pulled apart, giving details of all of the parts the object and how the parts fit together.

Fabrication - the act or process of forming and assembling materials and structures.

Finishing - the final stages of smoothing, painting, staining, and applying protective coatings to a project. Done both to enhance appearance and to protect the object from the environment.

Function - the use(s) to which an object or system is put.

Isometric drawing - a method of drawing that produces a three-dimensional view of an object. All vertical lines remain at 90° to the horizontal and all others are drawn at 30° to the horizontal. Isometric grid paper is recommended.

Interaction - the way that component parts act on each other or with each other.

Machine - a mechanism or device that helps people do work. The main machines are the lever, the pulley, the wheel and axle, and the inclined plane (wedge and screw).

Materials - the substance from which a structure is made.

Mechanism - the parts of structure that allow it to work or function.

Miter (mitre) box - a box used to hold a backsaw in a locked vertical position and while the cut is made in a piece of wood at a desired angle.

Model- typically a preliminary construction of some or all of a design, made from simple materials such as clay or paper, that can be used to test the design's features.

Modem - a device that connects one computer to another over a telephone line.

Orthographic drawing - a method of drawing the true shape of the surfaces of three-dimensional objects. Each surface is drawn separately as though being faced by the observer. Grid paper is recommended.

Power/energy - the resource that enables a mechanism to perform work.

Preference - any optional characteristics or factors that would be preferred in the solution to a problem. For example, a personal choice as to type of finish.

Prototype - the first practical try-out of anything. The original or model version.

Shear - where loads push at right angles to the surface of an object, such as when scissors are used to cut paper.

Spreadsheet - a piece of computer software that allows for the manipulation and display of numerical data.

Strut - The parts that have to resist compressive forces are called struts.

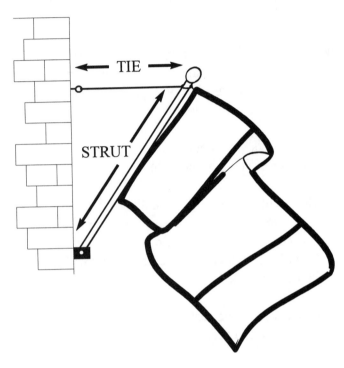

STRUT

TIE

Tension - where a load tends to pull or stretch an object apart, such as when two people pull on a rope. (In building construction, the parts that are in tension are call ties.)

Tie - In structural work, the parts that are in tension are called ties (ties are usually thinner than parts that have to resist compressive forces.)

Torsion - where a load tends to twist an object, such as when a person tightens the handle of a vice clamp.

Triangulation - where a triangular form is made in order to provide a strong rigid shape. For example, adding a strut made from card diagonally from one corner to another in a square shape, divides the square into two triangles, adding strength.

Structure - the essential physical or conceptual parts of buildings, etc. and the way in which they are constructed or organized.

System - comprehensive, self-sustaining combinations of interrelated structures, mechanisms, etc.

Technology - the use of knowledge or the practical means people use to change their surroundings.

Technotalk

Name: _____

Reaction to Technology

What does the word technology mean to you? In the wheel below, write or draw the first four things that come into your head when you hear the word *technology*. Exchange your wheel with someone else in your class. Compare your reactions.

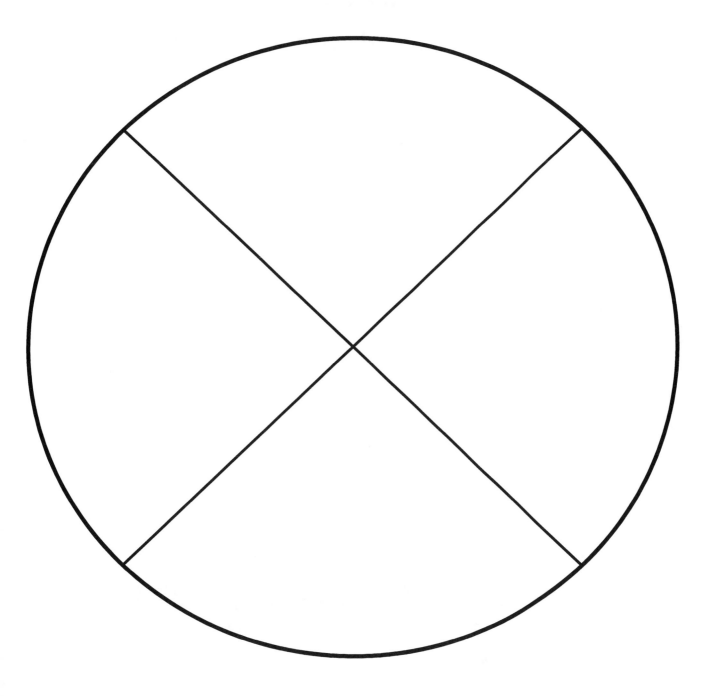

Technotalk

Self-assessment Form (K-3)

This is how I feel about my project.

Technotalk

Name: _____

Self-assessment Form (4-6)

Activity: _____

Circle the number that best describes your work. Think of it as a way of remembering how the project went and what you might do differently next time.

5	excellent
4	very good
3	good
2	satisfactory
1	poor

Effort

I tried as hard as I could.	1	2	3	4	5
I looked for help when I needed it.	1	2	3	4	5
I helped others in my group.	1	2	3	4	5

Investigating

I used books and other references.	1	2	3	4	5
I talked to people.	1	2	3	4	5
I used a computer search.	1	2	3	4	5

Construction

I contributed to the work.	1	2	3	4	5
I listened to others.	1	2	3	4	5
I had a task of my own.	1	2	3	4	5

Time

I did not waste my time.	1	2	3	4	5
I did not waste anyone else's time.	1	2	3	4	5
I finished on time.	1	2	3	4	5

Choose one item from the ones above in which you think you could improve. What would you do differently next time? _____

_____

Technotalk

Name: _____

Peer Assessment Form (4-6)

Activity: _____

Name of student being assessed: _____

Circle the number that best describes the student's work on the activity just completed. Try to be fair and honest.

5	excellent
4	very good
3	good
2	satisfactory
1	poor

Effort

Concentrated on the task.	1	2	3	4	5
Looked for help when I needed it.	1	2	3	4	5
Helped others in the group.	1	2	3	4	5

Investigating

Used books and other references.	1	2	3	4	5
Contacted people.	1	2	3	4	5
Used a computer search.	1	2	3	4	5

Construction

Contributed to the work.	1	2	3	4	5
Listened to others.	1	2	3	4	5
Took responsibility for a task.	1	2	3	4	5

Time

Did not waste my time.	1	2	3	4	5
Did not waste anyone else's time.	1	2	3	4	5
Finished on time.	1	2	3	4	5

Choose one item from the ones above in which you think this student could improve. Suggest what the student might do differently next time. _____

Technotalk

Name: _____

Group Assessment Form (4-6)

Activity: _____

Team Members: _____

Agree on your answers as a group. Circle your choice.

1. Did we share?	yes	sometimes	no
2. Did we take turns?	yes	sometimes	no
3. Did everyone contribute?	yes	sometimes	no
4. Did we listen to each other?	yes	sometimes	no
5. Did we help each other?	yes	sometimes	no

Finish each sentence.

6. We agreed on _____

_____

7. We disagreed on _____

_____

8. We each had a task. The tasks were:

_____

_____

_____

_____

9. We could improve by

_____

Assessment (K-3)

Student Name: _____

Date: _____

Activity: _____

Observed Criteria	Ranking (1 poor to 5 excellent)
confidence	
imagination	
practical skills	
organization	
leadership	
manual dexterity	
positive attitude	
math skills	
creativity	
flexibility	
reasoning ability	
co-operation	
planning ability	
motivation	
language skills	
hand-eye co-ordination	

Student Name: _____

Date: _____

Activity: _____

Assessment (4-6)

Observed Criteria	Ranking (1 poor to 5 excellent)

Reading List: Materials

Bang, Molly. *The Paper Crane*. New York: Mulberry Books. 1985. (This book uses Japanese paper to create origami including a paper crane.)

Evans, David and Williams, Claudette. *Building Things: Let's Explore Science*. Richmond Hill, Ontario: Scholastic Canada Ltd., 1993. (This book explores materials by asking questions such as: What is it made of? What is it like? What can you do to it? Is it strong or weak?)

Gibbons, Gail. *Up Goes the Skyscraper*. New York: Aladdin Books. 1986. (This book explores the materials and process involved in building skyscrapers from the initial stages at the drawing board to the actual finished product.)

Gilman, Phoebe. *Something From Nothing*. Richmond Hill, Ontario: North Winds Press. 1992. (This book transforms one piece of material from one thing to another. A tattered blanket becomes a jacket, then a vest, a tie, a handkerchief, a button, and finally a wonderful story.)

Graham, Thomas. *Mr. Bear's Chair*. New York: Dutton Children's Books. 1987. (Mr. Bear breaks his chair and has to go out in search of materials with which to fix it.)

Hutchins, Pat. *Changes, Changes*. New York: Aladdin Books. 1971. (This is a picture book of building blocks that change from a house to a fire engine to a boat, a truck, a train, and then back to a house.)

Morris, Ann. *Hats, Hats, Hats*. New York: Mulberry. 1989. (A book about all kinds of hats and their materials, such as straw for sunhats, fiberglass for bike helmets, etc.)

Provensen, Alice and Martin. *The Glorious Flight*. New York: Viking Penguin. 1983. (Recounts the life of Louis Bleriot and his desire to fly. Emphasizes the importance of selecting the right materials to build and aircraft.)

Williams, Karen Lynn. *Galimoto*. New York: Lothrop, Leo and Shepard Books. 1990. (A wonderfully creative use of wire – from nothing to a truck that becomes the envy of all.)

Wise Brown, Margaret. *The Important Book*. New York: Harper Trophy. 1994. (Although there are many ways to describe something, there is one specific important thing about it.)

Wynne-Jones, Tim. *Architect of the Moon*. Toronto: Meadow Mouse. 1988. (A brave block-builder receives a message that the moon is falling apart and needs his help.)

Reading List: Structures

Bardon, Keith. *Exploring Forces and Structures*. East Sussex: Wayland. 1991. (This text and activities are based on questions such as: Why do things move? What is drag? How can friction be useful?)

Drew, David. *The Paper Skyscraper*. Toronto: Ginn. 1992. (This is a book about the materials different things are made of, giving clues about a mystery item and questions what things would be like if made differently.)

Eisen, David. *Fun with Architecture*. New York: Metropolitan Museum of Art. 1992. (This is a kit containing architectural stamps and a book introducing basic architectural principles and styles from the ancient world to the present day.)

Eyewitness Visual Dictionaries. The Visual Dictionary of Buildings. Toronto: Stoddart. 1992. (This book looks at the inner structure of buildings of all kinds and from all time periods, ranging from castles to skyscrapers. It is also available in CD-ROM format.)

Isaacson, Philip M. *Round Buildings, Square Buildings, and Buildings That Wriggle Like a Fish*. New York: Alfred A. Knopf. 1988. (A global tour of buildings, discussing the many elements that give buildings their character, from materials to placement in the landscape.)

Jennings, Terry. *Structures. The Young Scientist Investigates*. Chicago: Children's Press. 1988. (Answers questions about why structures are shaped the way they are.)

Kaner, Etta. *Bridges*. Toronto: Kids Can Press. 1994. (This book looks at the main types of bridges and the science behind them. Includes current and historical information.)

Knapp, B.J. (ed.). *Shapes and Structures*. Toronto: Grolier Ltd., 1991. (Explores the way that basic geometrical shapes are translated into so many useful structures.)

Pollard, Jeanne. *Building Toothpick Bridges*. Palo Alta: Dale Seymour Publications. 1985. (A hands-on approach to learn concepts such as stress, fulcrums, gravity, and strength of various geometric forms.)

Steel, Sara. *Bridges*. London: Young Library. 1985. (This book contains dozens of amazing stories, pictures, facts, and ideas about bridges.)

Wilson, Forrest. *What it Feels Like to Be a Building*. Washington, D.C.: The Preservation Press. 1988. (Playful drawings and humorous text.)

Zubrowski, Bernie. *Messing Around with Drinking Straw Construction. A Children's Museum Activity Book*. Boston: Little, Brown, and Co. 1981.

Links with Self, School, Community, and Environment

Who can help you? A list of contacts can be one of your most useful resources.

Contact Name	Date Contacted	Role in the Class	Phone Number/ Address

Dear Parents/Guardians:

The students in our class will be going out on TECHNOWALKS. These are walks in the community to explore technology and how it affects people and the environment. We will begin with an exploration of materials (what objects are made of) and then look at structures (how objects are formed).

Aspects of technology will be introduced in the classroom prior to the actual TECHNOWALK and each walk will be followed up with activities back in the classroom and at home. We encourage you to share in this learning experience by discussing with your child the different activities we have been doing. We would also like to invite you to visit our classroom or to join us on our TECHNOWALKS at any time.

Thank you.

Please return the form below with your student.

• •

_ _ _ _ _ _ _ _ _ _ _ _ _ _ has my permission to leave school property for TECHNOWALK excursions with the class. I understand that such excursions involve risks and situations beyond the regular functioning of the school. I will remind my child of the importance of safe, responsible behavior during such excursions.

_____
Signature of Parent/Guardian

_____
Date

What is it made of?

How is it built?

What makes it go?

What is it used for?

How does it work?

How does it look?

Who helps make it?

Technotalk

Name: _____

CLASSIFYING MATERIALS

Record your materials along the top of this chart. Use a checkmark to record the properties of each material. For example, is it soft or hard? Flexible or rigid?

M A T E R I A L

PROPERTIES											
Soft											
Hard											
Flexible											
Rigid											
Heavy											
Light											
Elastic											
Inelastic											
Smooth											
Rough											
Dull											
Shiny											

Print Materials

Chapman, C., et al. *Collins Technology for Key Stage 3: Design and Technology, the Process.* 1992, London: Collins Educational. (Good ideas to help you with the design process.)

Corney, Bob & Dale, Norman. *Technology I.D.E.A.S.* Canada. 1992. Maxwell MacMillan. (Projects for elementary students.)

Corney, Bob, Reynolds, William, & Dale, Norman. *Imagineering – a "Yes, We Can!" Sourcebook for Early Technology Experiences.* Toronto. 1997 Trifolium Books Inc. (Activities and teacher support, Grades K-3)

Metropolitan Toronto School Board. *All Aboard! Cross-Curricular Design and Technology Strategies and Activities*, Toronto. 1996. Trifolium Books Inc. (Theme-based activities and teacher support, Grades K-6.)

Metropolitan Toronto School Board. *By Design, Technology Exploration & Integration*, Toronto. 1996. Trifolium Books Inc. (Problem-solving through design and technology. Activities and teacher support, Grades 6-9.)

Peel Board of Education Teachers *Mathematics, Science, & Technology Connections*. Toronto. 1996 Trifolium Books Inc. (Activities and planning assistance, Grades 6-9)

Richards, Roy. *An Early Start to Technology.* London. 1990. Simon & Schuster. (Emphasis on things children can readily design and make through problem-solving.)

Recommended Construction Kits

Grades K-3

Duplo. (Larger version of Lego designed for young children. Excellent introduction to Lego. Permits imaginative play.)

Lasy (Imaginit). (Plastic rectangles, connectors, etc. Versatile.)

Stickle Bricks. (Squares, triangles, and other pieces which "stick" together. Can be combined with "straws." Easy to use.)

Grades 4-8

Baufix. (Wood and plastic, similar to Meccano. Students uses nuts, bolts, and spanners. Durable and colourful.)

All Grades

Lego. (Very versatile. Suggest transferring the pieces in a large box with many small compartments, rather than the original cardboard box. Some components are fragile.)

Meccano. (Original version can be difficult for younger students, but there is a new set with plastic that is easier for them to use.)

Ramagon. (Used by NASA engineers to model space stations. Easy to use and works with other construction kits. Durable.)

Some Teacher Resources

Some Internet Sites

The following Yahoo search page is specifically of interest to students and teachers. It allows users to search for information using keywords, but confines its search to things that might be of interest to educators.[2]

http://beta.yahoo.com/Education/K_12

The following web site, posted by *Discover Magazine*, is full of links to other science and technology related sites. It is an ideal research starting point for students.

http://www.enews.com:80/magazines/discover/

Sites about Math

MathPro: puzzles, problems and resources[3]

http://sashimi.wwa.com:80/math/

The MegaMath Glossary and Reference Section. [3]

http://www.c3.lanl.gov/mega-math/gloss/gloss.html

Fun Math Things. Includes: paradoxes and logic puzzles. [3]

http://www.uni.uiuc.edu/departments/math/glazer/fun_math.html

Other Sites of Interest

A learning kit of activities and lesson plans focused on motion.[3]

http://schoolnet2.carleton.ca/english/worldinmotion/index.html

Elementary and Junior High. A young student describes 3 successful projects that could be used as activities. Erosion, Battery Power, Soil Pollution. [3]

http://megamach.portage.net:80/~bgidzak/nick.html

Australia: A girls' science and technology high school has a webpage with links to some of their own experiments and projects in robotics and other things.[2]

http://www.ozemail.com.au/~mghslib/projects/mghsproj.html

An American webpage similar to the Australian one, above, that has good ideas, can be found at this address.[2]

http://forum.swarthmore.edu/sum95/projects.html

> **To Go Further**
> We highly recommend you use the resource, *The Teacher's Complete & Easy Guide to the Internet*, Heide, Ann & Stilborne, Linda, 1996 Trifolium Books Inc., Toronto, Ontario ISBN 1-895579-85-6 both as a source of thousands of tested sites, and also as a guide. Find out more about this and other resources by visiting Trifolium's website, www.pubcouncil.ca/trifolium

---

[1] Thanks to Professor Don Galbraith for allowing us to use his introduction to Netscape here (adapted from *A Portfolio of Teaching Ideas for High School Biology*, Trifolium Books Inc.)

[2] Thanks to John Rising for locating sites he thought would be of special interest.

[3] Thanks to Ann Heide and Linda Stilborne, *The Teacher's Complete & Easy Guide to the Internet* for allowing us to include a few of the many sites they have identified

More About Trifolium Books

Trifolium Books Inc. is a publishing house that specializes in teacher and student resources. One of its mandates is to search out and publish exemplary materials produced by teachers. It is Trifolium's hope that you will find some or all of the following resources helpful in your classroom.

Professional Resources

For Those Interested in the Internet and Distance Learning

The Teacher's Complete and Easy Guide to the Internet, Heide, Ann & Stilborne, Linda (Comprehensive and educator-specific resources, with activities, tested Internet addresses, active website support, and more.)

Internet Handbook for Writers, Researchers, and Journalists, McGuire, Mary, Stilborne, Linda, McAdams, Melinda, & Hyatt, Laurel. (Covers research and writing from search techniques to copyright issues online.)

Why the Information Highway? Lessons from Open & Distance Learning, edited by Roberts, Judith & Keough, Erin (Practical applications supported by pertinent theory and analysis, presented by colleagues established in the use of the latest technologies.)

For Those Using Computers and Other Information Technologies in Class

The Technological Classroom: A Blueprint for Success, Heide, Ann & Henderson, Dale (A practical guide to integrating technology in all educational settings, from floor plans to strategies for active learning.)

For Counselors and Child Care Workers

Beastie Girls, Betties, & Bangers: Violence and the Adolescent School Girl, Artz, Sibylle. (The life-worlds and practices of girls who are involved in violence, but not yet involved with the justice system. Includes case studies.)

Feeling As a Way of Knowing, Artz, Sibylle (Explores the importance, meaning, and value of emotion in understanding oneself and the world. Offers a unique six-step strategy for unravelling the meanings behind specific emotional experiences.)

Other Resources Suited to Grades K-8

Career Connections Series, Trifolium Books Inc. and Weigl Educational Publishers. (Career exploration integrated with activities in math, science, technology, and other subject areas. Each of the 18 student books focuses on a particular area of interest, such as the outdoors, working with people, or working with your hands, and consists of illustrated interviews with 10 real people about their careers. Three Teacher Resource Banks provide additional activities and teaching strategies.)

All Aboard! Cross-Curricular Design and Technology Strategies and Activities, The Metropolitan Toronto School Board (Theme-based activities and teacher support, Grades K-6.)

By Design, Technology Exploration & Integration, The Metropolitan Toronto School Board (Problem-solving through design and technology. Activities and teacher support, Grades 6-9.)

Imagineering – a "Yes, We Can!" Sourcebook for Early Technology Experiences, Corney, Bob, Reynolds, William, & Dale, Norman. (Activities and teacher support, Grades K-3)

Mathematics, Science, & Technology Connections, Peel Board of Education Teachers (Cross-curricular activities and planning assistance, Grades 6-9)

If you wish further information about these or our other titles, please contact: Trifolium Books Inc. 238 Davenport Rd., Suite 28, Toronto, Ontario M5R 1J6; Tel: (416) 925-0765; Fax: (416) 485-5563;

Or check out our website: www.pubcouncil.ca/trifolium

Dear Reader,

Has this book "worked" for you? We hope you have enjoyed using the ideas in *Take a Technowalk to Learn about Materials and Structures*, and feel that by using these ideas, you have enhanced your students' problem-solving skills <u>and</u> desire to learn. We are pleased to have the opportunity to make this book available to you, and we hope you will find other resources in our **Springboards for Teaching** series, as well as Trifolium's other educational resources, of great value both to your students and for your own learning.

None of us are immune to the praise and criticism of others, and those of us who write, edit, and publish educational resources such as this one are no different. Please let us hear from you about, yes, both your criticism and your praise. We need to know what you find useful about each of our resources and what you would like to see developed differently. We are committed to producing exemplary resources that are truly useful to you. Thus, if you have any comments or recommendations that you feel would assist us in the development of future projects and perhaps a new edition of *Take a Technowalk* in the future, we welcome them. Please write, fax, or e-mail Trifolium at your convenience.

We have found that the concept of TECHNOWALKS has stimulated other teachers to come up with their own versions. We would be very pleased to hear from you about how you may have taken this approach in different directions. For example, have you found any new activities to use before or after? Are there any places in your community that have proven particularly worthwhile? Please also send us any comments from your students, parents, or yourself about TECHNOWALKS. We are determined to continue to develop and improve the material we offer to you and your fellow educators.

Do you have a project idea that you think other teachers would find helpful? If so, please send it to us for consideration. Our aim is to continue providing practical, effective educational resources, particularly in the areas of mathematics, science, and technology, and also in career development.

We look forward to hearing from you with your thoughts about this book, new project ideas, or both!

Yours sincerely,

Trudy L. Rising
President

Trifolium Books Inc.
238 Davenport Road, Suite 28
Toronto, Ontario M5R 1J6
fax 416-485-5563
www.pubcouncil.ca/trifolium